PERSON
EVO

by the same author

TOTAL MAN
Notes Towards an Evolutionary Theory of Personality

(*with M. L. Kellmer Pringle*)
FOUR YEARS ON
A Follow-Up Study at School Leaving Age of
Children Formerly Attending a Progressive
and a Traditional Junior School

PERSONALITY AND EVOLUTION

The Biology of the Divided Self

STAN GOOCH

WILDWOOD HOUSE
LONDON

First published 1973
© 1973 by Stan Gooch
Wildwood House Ltd, 1 Wardour Street, London W1V 3HE
ISBN hardback 0 7045 0047 7
ISBN paperback 0 7045 0048 5

The quotations in Chapter X from Schiller's *On the Aesthetic Education of Man,* edited and translated by Elizabeth M. Wilkinson and L. A. Willoughby (Oxford University Press, 1967), © 1967 by Oxford University Press, are reprinted by kind permission of The Clarendon Press, Oxford.

Condition of Sale

This book is sold subject to the condition that it shall not, by way of trade or otherwise, be lent, re-sold, hired out, or otherwise circulated without the publisher's prior consent, in any form of binding or cover other than that in which it is published and without a similar condition including this condition being imposed on the subsequent purchaser.

Printed at the Compton Press Ltd.,
Compton Chamberlayne, Salisbury,
Wiltshire.

Contents

Preface	1
I. THE BACKGROUND	5
II. PREVAILING WINDS OF THE NERVOUS SYSTEM	
1 Archetype and Archestructure	47
2 Ego-Archetypes and Ego-Archestructures in Western Civilization	60
III. BIOLOGY AS DESTINY	
3 Body-Types and Personality	83
IV. THE DECLINE AND FALL OF BIOLOGY	
4 The Energy of Evolution	93
5 Darwinism: A Partial Equation	105
6 Symbolic Evolution	116
V. THE INWARD UNIVERSE	
7 The Nature of Consciousness	135
8 'On the Aesthetic Education of Man'	153
9 True and False Alternative Societies	169
10 Beyond the Archestructure	186
VI. EXPERIMENT AND VALIDATION	197
Note on the relationship of libido and aggression	204
Bibliography	207
Index	209

Preface

The present book, from one point of view, continues the exposition of a theory of personality outlined by myself in an earlier book, *Total Man*.[1] An introductory section in this new book, therefore, summarizes the main argument of that previous book, since some prior acquaintance with the general theory involved is advisable. Where in fact I refer to the idea of the archestructure, or the concepts of Self, Ego and Person, a prior acquaintance is probably indispensable. On the other hand, the discussion and definition in the present book of the 'I' (the conscious element of our personality which seems somehow able to move across and within the personality dimensions referred to) is not only new, but can in any case stand by itself quite independently of my particular theoretical basis. Anyone already familiar with my previous book may certainly if so inclined begin with Chapter 1. The introductory section does nevertheless contain a fair amount of new detail not included in *Total Man*.

A second justification for a relatively extensive preamble arises from my wish to give fellow psychologists and social scientists as clear and summary a statement as possible of my general theory. Since I have been at times extremely critical of the views of many professional colleagues, it is only fair that I should now offer them an unequivocal statement of my own position, in order that they in turn may criticize that.

While, then, I have certainly had social science and the social scientist very much in mind throughout the whole of the present book, it is, nonetheless, so written that the generally informed reader will be able to follow both the text and the development of the

---

[1] (Allen Lane, London, 1972).

arguments without difficulty. The majority of matters I discuss, though in some sense 'scientific', do in most cases also form part of our general cultural and educational heritage. Partly for this reason, I have felt justified in largely dispensing with the dead weight of detailed references which usually make up a significant portion of any scientific text. Where called for, however, I have provided detailed reference to particular sources, or otherwise indicated a general debt to a specific discipline.

One of the further aims of the present book is to demonstrate that classical Darwinism, despite its usefulness, is no adequate model, because an incomplete one, on which to base an understanding and appreciation of evolution – much less, then, of the psychology of organisms; least of all of our own subjective, experiential lives. This statement should not at all be taken to indicate that I am any kind of enemy of the evolutionist or the geneticist. On the contrary, my position here will be felt by many, in these environmentally persuaded times, to be thoroughly reactionary. For it seems clear to me, for example, that political allegiance itself is a characteristic inherited along normal Mendelian lines.

Unlikely as it must seem after that comment, however, my main intention in this book is to bring inspiration to those who seek to question the academically fashionable, mechanistic view of the organism, at both the biological and psychological levels. I would hope very much to have supplied a good deal of ammunition for use in the face of denials from the mechanistic/materialist position by its nature much more easily – that is, more glibly – argued.

As my general terms of reference, in particular the constructs of Self, Ego and Person, become more familiar, I hope that what seems to me their great benefit will be recognized – namely, that with them we are in a position to discuss very large and apparently unconnected areas of human activity within one unifying framework. Instead of then speaking of, say, literature, religion, politics, economics or physics, we find ourselves speaking instead only of different aspects of man's psychological being – for this is all in the last analysis that these activities are, or in fact can be. By the application of my suggested terms we are, for example, at once aware of the psychological content of political statements or the psychological infrastructure of theoretical physics. This, if indeed the case, seems to me to represent a very considerable gain.

I
THE BACKGROUND

The Background

In my earlier book, *Total Man,* I outlined and defended a general theory of human personality based on evolutionary and developmental considerations. I attempt to convey the nub of that theory here in summary form before moving on to the considerations of this book proper.

Theories of personality try to account for and incorporate the manifold moods and behaviours of the individual within the framework of some proposed overall structure or underlying dynamic; to provide a theoretical platform, as it were, from which to debate, and hopefully account for, the many complexities and seeming contradictions of the human personality. My own particular definition of 'personality', I must emphasize, is one that embraces all mental contents – the total psychological life of man. The term as I use it has the same all-encompassing significance for mental life as the term 'body' has in respect of our physical attributes. By no means all writers, however, use the word personality in so wide a sense, many preferring to apply it only to the more emotional or affective aspects of mental life – tending to place intellectual activity and cognition, for example, in another compartment. I regard such compartmentalization of our psychological life rather as a comment on the personality type of those performing this operation than as a useful way of approaching the human psyche. I insist upon this position regardless of the great benefits which compartmentalizing approaches have undoubtedly conferred in the physical sciences. As I consider it, this approach is itself a behaviour – as of course is everything that we do – and is therefore among the quantities of information to be taken into account when attempting to assess the nature of personality. It is, indeed, one of my basic assumptions that *all observable phenomena,* both those of our own lives and those of the universe, whatever their other values, functions or significances in any local context, are clues to and, in one sense or another statements about, the nature of human personality and the human nervous system.

Turning therewith to the theory itself, I have proposed that the essential basis of the personality is dual or double – that is, I consider it to be made up of two major, in many ways separate, parts. In the earlier book I broached this position in the first instance by inspecting and discussing a large variety of cultural phenomena, mainly, though not wholly, drawn from the European arena, both past and present, which over and over again seem to show in symbolic and artistic form the picture of a divided personality. These include all versions of duality and dualism in religion and philosophy: the preoccupation with opposites, oppositeness and polarity in virtually all times and cultures (black–white, East–West, man–woman, God–Devil, and so on); the preoccupation (in stories, plays, legends) with the double or twin, with mistaken identity, and with one's own reflection; and the further preoccupation with puppets, dolls and mechanical toys. Added to these examples are all stories of ghosts and magical and diabolical possession, especially of such figures as the Follower and the Shadow. Finally, in what are at first apparently quite unrelated fields, the political divisions of Left and Right; the two major forms of basic learning, classical and operant conditioning; and the two major groups of psychiatric illness, neurosis and psychosis. All instances so far given are derived from the psychological-sociological stratum of behavioural phenomena. At the biological-evolutionary level, however, a further extremely important instance of the dual personality is the sexual division (of all but the lowest animal species) into male and female.

In terms of my theory of personality, to revert to the stricter area of professional psychology, I term these two major components of personality the 'Ego' and the 'Self'.[1] The first of these is an already well-established construct both within and beyond the specialist literature of psychology. The affective and motivational qualities I assign to it are, further, in general well agreed among psychologists – assertiveness, aggressiveness, intentionality (or will), and so on. Many psychologists and psychoanalysts, among them of course Freud, additionally assign to the ego the major attribute of consciousness, as well as a close association with material reality. In this I also concur. It is probably in my further explicit assigning of

---

[1] Throughout the whole of this book I write 'Self' for my own construct, and 'self' with a small 's' when referring to the constructs of other psychologists; and similarly 'Ego' with a capital 'E' when referring specifically to my own concept.

logical thought, rationality, cognition, problem-solving and similar related processes and behaviours to this, now much broader, personality construct that I part company with the main body of personality theory and, as many would feel, cross a boundary into other territories. Thus briefly stated here, there may indeed seem many instant objections to my position. A good number might feel, for example, that aggressiveness and rationality, two of the attributes I have assigned to the Ego, make very strange bedfellows. Surely rational man is not aggressive? I shall, however, not take up these apparent difficulties at once – on promise of showing a little later that no paradox or loose thinking is involved.

The word 'self' also already exists as a term in extensive use in the literature. My definition of the Self is, however, new. Most psychologists employ 'self' more or less as a synonym for 'ego' – although the term then does often carry a suggestion of a more inward aspect of personality. The fact that there should be *two* terms in the psychological literature is, again, of considerable significance, suggesting that psychologists are unconsciously aware of basic duality. C. G. Jung, somewhat exceptionally, uses the term self to refer to the *totality* of the personality – ego plus the rest. In my own theory, the Self is the shadow or 'female' version of the Ego: an elusive, but equal, partner (indeed, in many of the senses of that term a 'sleeping partner') in the affairs of the total personality. Though in many ways apparently 'weaker', in yet other ways the Self is actually the stronger partner. This is partly a matter of what activities and which particular parts of the twenty-four-hour day/night cycle are in question. The most important single point to grasp in connection with the Self – and with this one insight we suddenly move a very long way forwards in our understanding of its nature – is that the Self's attributes are in every case and in every way *opposite* and *antithetical* to those of the Ego. This will become very evident.

The Self as a persona has the same relation to (or the same function within) what is nowadays often termed the 'unconscious' as the Ego has to, and within, the conscious sector of personality. The Self, we can say, is the *personification* of the unconscious. With this said, I must nevertheless emphasize that the term unconscious is not one that I particularly favour. I use it because it is after all a well-established and therefore convenient referent. I myself view the unconscious very emphatically as another form of *consciousness*. This I therefore frequently refer to as 'night' or 'sleeping' consciousness. When night consciousness is in operation – e.g. when we are dreaming – normal or 'day' consciousness is itself at that time as 'unconscious' (i.e. as out of consciousness) as the so-called unconscious is during our waking hours.

If the Self is the 'female' version of the Ego, and if the term female, as yet undefined, nonetheless bears at least some of its normal associations, it should follow that the Self will be more in evidence, or be more dominant, in women than in men. Such is the case. The converse then should and does apply respecting males and the Ego. Here I depart from the view of C. G. Jung, who maintained that in the male the female aspect of the personality resides in the unconscious, while in the female the male element so resides. My own view maintains that what I term the Self is lodged in the unconscious in *both* male and female; while the Ego of each resides in consciousness. On the separate point that women are relatively more governed by their unconscious (by night consciousness) than are men, I am, however, in line with the psychoanalytic movement as a whole.

Ego and Self, then, are the two personifications or expressions at the purely psychological level of the basically bipartite nature of personality. When referring, however, to the two mental 'universes' in which these two personae operate, or when referring to the dynamics or laws of those universes, I usually speak instead of System A (= Ego) and System B (= Self) respectively. These systems are on the one hand still *psychological* constructs and refer to psychological events. However, we have now moved away from the actual *personifications* (the us-as-Ego, us-as-Self *experience*) of the Ego and Self already in the sense that not all the events of Systems A and B are necessarily available to the appropriate consciousness. For example, we are in many ways – and sometimes exclusively – aware rather of the results or outcome of thought processes than of the actual detailed thought processes themselves.

In the concepts of Systems A and B we have also moved appreciably closer to that nebulous and difficult line which separates the psychological (or mental) from the physiological (or purely physical). With Systems A and B we are referring much of the time not to conscious awareness (though that too), but to the mental or psychological correlates of observable, or at least hypothesized, processes, taking place at the physiological and biochemical levels.

Systems A and B, as already suggested, may be thought of in one way as the 'universes' which the Ego and Self respectively inhabit. I have proposed that these two universes are paralleled, and in some sense underpinned at the physiological level by the so-called *central* and *autonomic* divisions of the physical nervous system.[2] These

---

[2] One is of course far from saying here that the psychological and the physiological are one and the same. Each certainly exists, and perhaps can only exist, in association with the other. A possibly useful analogy here is that of the magnetic field which forms about a wire carrying an elec-

physiological divisions are well agreed in the specialist literature. (There are, I must emphasize, still other ways of dividing up the physical nervous system, the details of which need not concern us for present purposes. For example, the term 'peripheral nervous system' is used, and refers to the nerve complexes which carry messages between the central nervous system and the autonomic system, and their respective receptors and effectors. These nerve complexes are, therefore, merely adjuncts to the c.n.s. and the a.n.s. and not themselves initiating or 'commanding' systems in the true sense of those terms. It is solely with *initiating* systems that I am concerned. I should finally perhaps mention that any implied watertight discreteness between the major systems we are discussing is unintentional. In fact and in practice all systems of the body interact, and sometimes overlap, rather complexly.)

The *central* nervous system (or the cerebro-spinal system as it is sometimes also called) is principally concerned with the skeletal muscles, with volitional motor movement, and with various of the external receptors – eyes, ears, nose and so forth. The *autonomic* system is notably concerned with internal glands and organs, with the smooth or involuntary muscles of the heart, viscera and so on, and with general economic factors such as the regulation of body temperature, growth, digestion and excretion. As a generalization, the central nervous system may be said to be concerned with the relationship of the organism to the *external* world, and the autonomic system with the management of the organism's *internal* environment.

The term 'autonomic' actually means 'self-governing'. This is an

---

trical current. The field is not the current, but nevertheless depends wholly on that flow for its existence. In terms of this analogy one might say that mental or experiential life is a 'field' surrounding a living brain (perhaps in a rudimentary sense every living cell). While there are aspects of consciousness in respect of which this analogy cannot serve us, it is by no means an inadequate point of departure.

There is also a rather different point involved here – the question of how far a psychological theory *requires* a physiological infrastructure to achieve validity. My view is that in principle a psychological theory (like, for instance, a chemical or mathematical theory) *need* only be valid in its own terms of reference. However, in practice, and for a wide variety of reasons which I do not wish to raise here, I consider it of the greatest importance that a physiological infrastructure be in general demonstrable. The issue is a complicated one, because psychology and physiology are differing levels of explanation, and the former can never be finally reduced to the latter. But the clash of psychological and physiological explanations is, nevertheless, always to be regretted – and I think inevitably indicates that one or other theory (or both) is in some error.

interesting description, which might lead one to ask where the seat of such government might be. The academic psychologist seems not so led – or, rather, he tends to assign control of the autonomic system to sub-cortical centres, such as the hypothalamus. Yet in fact these seem to be not more than co-ordinating centres. To consider them as executives is, in my opinion, to confuse the computer with its operator (a widespread delusion), whereas the former is and can only be the tool of the latter. I, personally, find myself obliged to look elsewhere for an alternative seat of government – like, and at the same time unlike, C. G. Jung. Jung, while similarly insisting on real and not simply notional self-government for the autonomic system, *was* apparently prepared to settle for the sub-cortical centres.[3] I am not.

We turn now, however, to another aspect of the nervous system – the anatomical structure of the brain itself. Some views of the human brain, both entire and in section, together with the stylized brains of a pigeon, dog and gorilla, are shown in Figures 1a and b. We are concerned especially with the *cerebrum* and the *cerebellum*.[4]

The fore-brain, which makes up the whole of the visible outer brain, is the seat certainly of waking consciousness, also apparently of memory, and of the higher, more typically human skills and abilities. The function of the hind-brain, as both psychologists and physiologists concede, is in general less well understood. For our present purposes, its pertinent features are these. First, its relatively large size. Second, the fact that, while an extremely ancient organ biologically (reaching its peak of *relative* development already in reptiles and birds), the cerebellum reaches its *absolute* peak of development in man. This fact alone might lead one to suppose it to play some major role, for in nature organs do not continue to evolve unless concerned with present events. Third, the style of its deeply convoluted outer surface, which so much recalls the surface of the cerebrum – and indeed, the word cerebellum literally means 'little cerebrum'. Further, the fact that the cerebellum has (in all mammals) its own independent connections with all parts of the nervous system below the medulla, as well as then further complex links with the cerebrum itself. Why, one might ask, given the complex links

---

[3] See *Collected Works of C. G. Jung* (Routledge, London, 1969), Vol. 8, pp. 509-10 and 510-11.

[4] In the developing human embryo we find at an early stage a *three*-part brain – consisting of fore-brain, mid-brain and hind-brain. In the finished brain, however, the mid-brain is more or less vestigial (although still not without its importance), a link between fore- and hind-brains, if anything more closely associated with the latter. The fore-brain and hind-brain, however, have now become the *two* major co-ordinating and directing centres of the nervous system, cerebrum and cerebellum respectively.

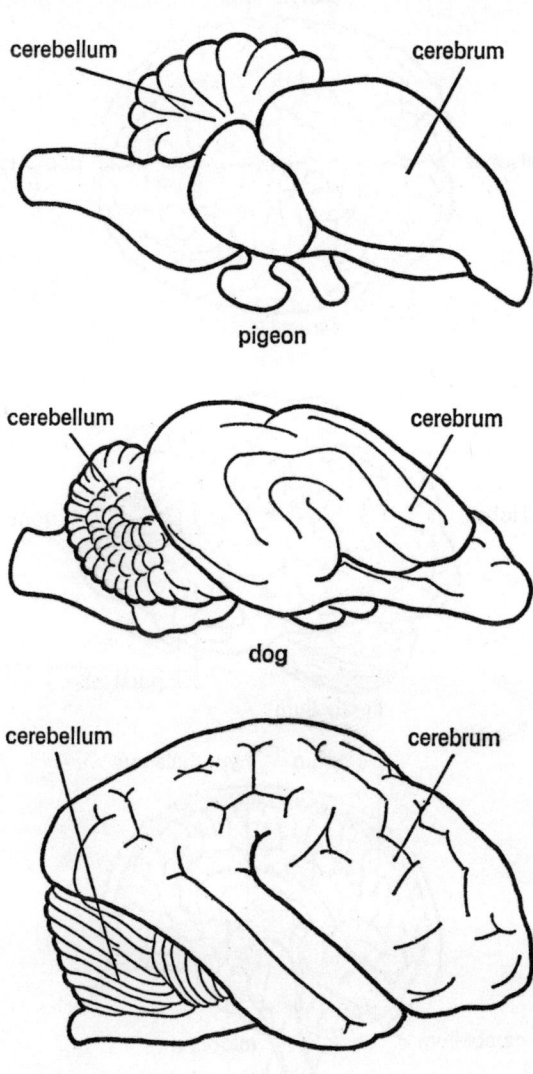

FIGURE 1
(a) External brain of the pigeon, dog and gorilla.

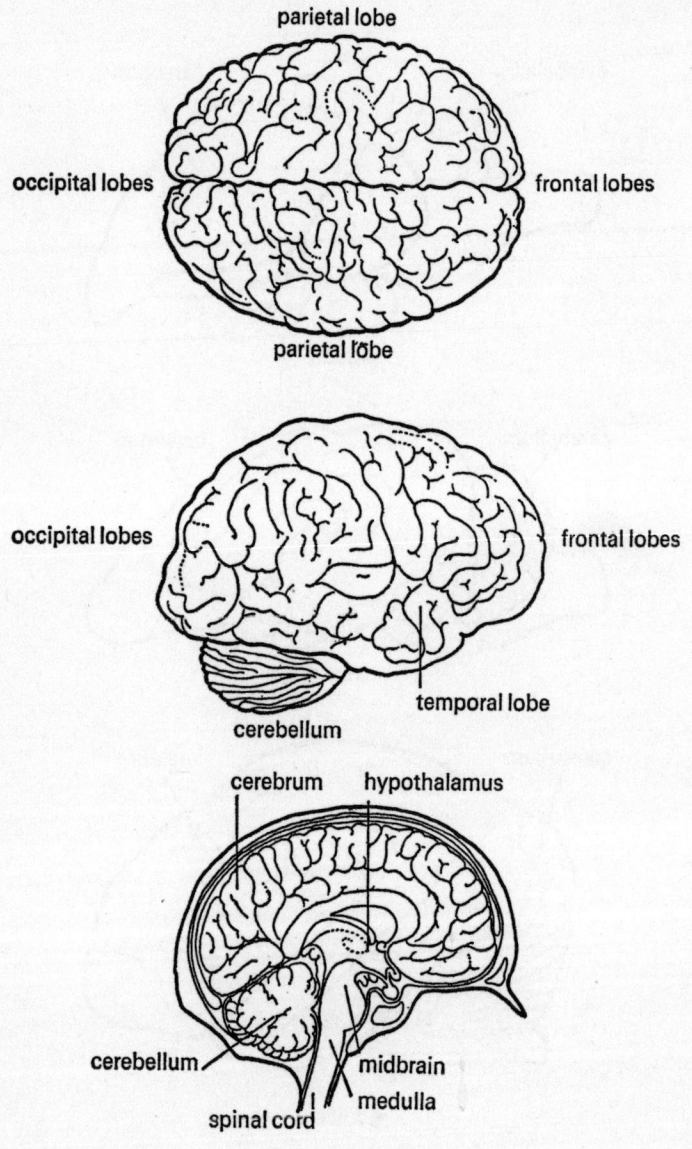

(b) Human external brain viewed from above and right: and internally in lateral section.

with the cerebrum, should the cerebellum additionally have its own independent connections with lower centres – unless, perhaps, in the interests of self-government?

From such circumstantial evidence, along with a wide variety of other direct and tangential evidence, I have sought to show that the cerebellum is somehow the principal seat of alternative consciousness – that is, of the Self. I am inclined to believe both that the cerebellum has some kind of consciousness of its own, but certainly and perhaps more especially that it can also make use of and generate consciousness in cerebral centres, and influence *their* general function – more strongly, for example, during dreams, but also during waking consciousness – hence the Freudian slip and other motivated 'errors'. During dreaming time, then, the *cerebrum* serves the *cerebellum* as an auxiliary, a reversal of normal waking conditions.

The cerebellum is, I suggest, a kind of brain within the brain – and yet more, an organism within the organism. The exploded, stylized mammalian brain of Figure 2 shows that this cerebellar 'organism' possesses, at least in rudimentary form, all features necessary for an individual existence – for example, primitive audial and visual centres, and the pineal body, an organ not merely once sensitive to light, but one which in certain reptiles apparently actually existed as a second pair of eyes. Though this point seems generally conceded in the specialist literature, I have personally never seen a drawing of a reconstructed reptile with four eyes. This is perhaps

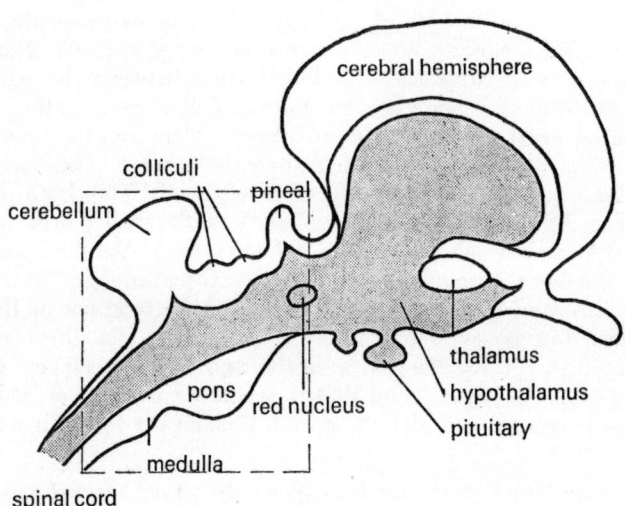

FIGURE 2
'Exploded' stylized mammalian brain.

one small example of conscious man's ramified denial of the Self.

The cerebellum, though large in its own right, is nonetheless small relative to the cerebrum. Moreover, in the course of evolutionary time, it has become covered by the ever-developing, backward-extending cerebrum – a process, I suggest, further encouraged by the adoption of an upright stance by our early ancestors. The progression described can be partially confirmed from the series in Figure 1a.

I believe, and have tried to show, that a large part of the magical, religious and narrative legends of Europe and the Near East are in fact actually concerned with the cerebellum and its relation to the cerebrum, although in indirect and symbolic, not literal, terms, and possibly without even any conscious awareness of this fact. Anything in legend or story which is any or all of the following is, I suggest, actually an unconscious reference to the cerebellum: small, lost, buried, different, alien, magical. Actual examples are: fairy-land, the Garden of Eden, Atlantis, the moon, the Holy Grail, Aladdin's cave and Aladdin's lamp, the dwarf, leprechaun and their numerous cousins (in particular the hump of the hunch-backed dwarf, and therefore Punch), the Old Man of the Sea, the baby Moses hidden in the rushes – and so on, almost indefinitely. The ghosts, devils, apparitions and the phenomena of magical possession already referred to also relate to the cerebellum in that these are, as I suggest, personifications, or projections, of one aspect or another of the Self.

Clearly this is a wide spectrum of examples. It needs to be recalled, however, that each of the major divisions of personality we are discussing is nothing less than an experiential *universe*. Nor, in any case, are the differences and disparities between the various symbolic forms of either the cerebellum or Self as great as they first seem, or as great as the differences between them and the symbolic forms of the Ego – a matter to which we shall return. The symbolic cerebellum, for instance, is always *relatively* small. Thus both fairy-land and Atlantis are small in relation to the real world (even though Atlantis is absolutely much bigger than Aladdin's lamp), just as the dwarf is small in relation to a real person.

The nature of the Self is, however, best understood not by direct reference, but by reference instead to the Ego, for three main reasons: first, because the Self is elusive and not given to revealing itself in broad daylight – and that is, whenever the Ego or waking consciousness is up and about;[5] second, because the Self is in many

---

[5] The Loch Ness Monster, incidentally, whether or not any such creature actually exists, is an excellent instance of the symbolic representation of the Self and the cerebellum – the prehistoric, reptilian nature of the monster referring clearly to the ancient origins of the organ.

ways (though not all) the 'weaker vessel' – is, as it were, less well integrated, even less well formed, in an evolutionary sense; third and most importantly, however, because the characteristics of the Self, as far as we can readily identify them, turn out *always* to be the precise *opposite* of the attributes of the Ego. It is therefore possible without difficulty to produce a fairly detailed picture of the elusive Self simply by turning around each and every quality of the Ego.

It is this last fact, together with our ability somehow to perceive or know the fact without having been taught it, which I believe further accounts for our particular interest in mirrors and in our reflection, and the role of these in fairy-tale and legend, as also with paired or linked opposites of all kinds. This, too, is ultimately why the Devil in mediaeval portraits steps from the magic circle with his left hand oustretched. For when we look at ourselves in a mirror our hands, so to speak, change places. When we move our right hand, the image in the mirror moves its left. If one imagines instead of the reflection in the mirror two real people facing each other, the notion involved becomes clearer. (A small, related point here is that the vampire and the ghost in legend have no reflection, because, I suggest, they *are* that reflection – that is, are manifestations of the Self, the mirror-image of the Ego.)

A further fact of cerebellar life is also included in this situation. For messages ascending from the lower organs and receptors of the body from one or other side cross over, just before reception, to reach the *opposite* side of the *cerebrum* – not the same side, as one might expect. Messages to the *cerebellum*, however, do not cross and are relayed to the same-sided, cerebellar hemisphere. Direct communication between the cerebellum and the cerebrum then of course necessitates a further crossing, or re-crossing, of nerve-fibres to cancel out the result of the earlier cross. (From one or both of these circumstances arises, as I believe, the widespread symbolic appeal of the cross – Christian, crooked and otherwise.) Thus when cerebrum and cerebellum 'face' each other the same switch has occurred as occurs when we look at ourselves in a mirror (see Figure 3). The longstanding evil reputation of the left (and perhaps some actual left-handedness) is, I believe, rooted in this situation.

Given that the nature of the Self is always opposite to the nature of the Ego, we were able to say at once that the Self is irrational, illogical, non-competitive, non-assertive, and so forth – i.e. the reverse of the many characteristics assigned to the former. There is a problem here, however – one which we shall avoid for the moment – and that is that the terms we have just employed describe the Self only *from the Ego's point of view*. The terms which the Self would, and sometimes does, use to indicate its own qualities certainly do

(a) two individuals facing each other viewed from above

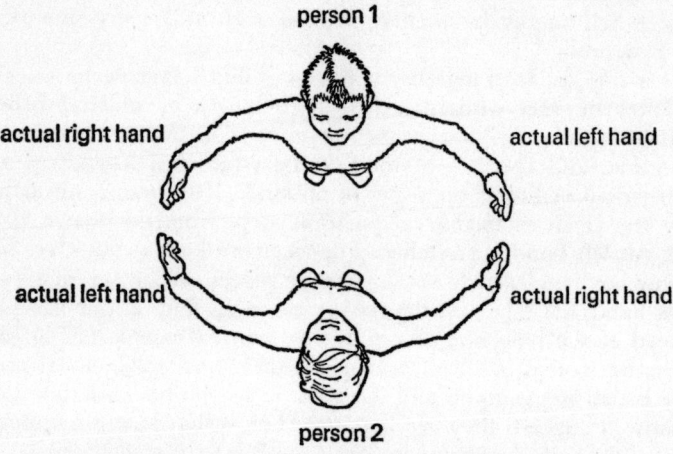

(b) one individual looking at himself in a mirror, viewed from above

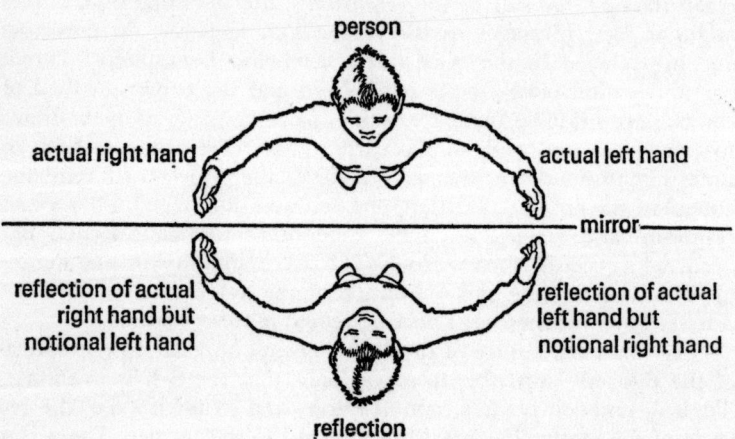

FIGURE 3
Mirror image and handedness.

not have the pejorative overtones we immediately detect in the Ego's descriptions. Thus, for instance, 'non-assertive' might be termed 'receptive', 'illogical' termed 'intuitive', and so on.

The general reversal I have suggested of attributes like logical and assertive is perhaps not a notion over-difficult for anyone – psychologist or otherwise – to accept. It is when we take the very laws of the universe – that is, the laws of the material, physical world which we perceive and inhabit during waking consciousness – and reverse *those* that a very interesting and, indeed, very disturbing area is entered. This is none other than the world of magic.

Such is the dominance of the Ego in our consciousness and in the scientific and academic western world that the notion that the 'laws of the universe' are at best the laws of our interaction with it, and are indeed often little more than aspects of our own consciousness, seems in itself somehow an affront. With this, incidentally, I do not seek to deny that some kind of physical universe exists in its own right, and therefore of course in our absence. Nonetheless, all perception involves a means of perception, and the properties of what is perceived are, inevitably, always partly a function of that means.

There is a further aspect involved, this being that the probing, cataloguing Ego in the first place actually evolved in close and direct contact with the external universe – or, to use Freud's expression, the Reality Principle. The Ego, therefore, cannot *help* but bear the former's stamp, and have incorporated the former's nature in the very heart of its own being (or, more precisely, in the head of its own doing). The laws which the Ego (that is, we) discovers in the universe *are the laws which the universe has pre-formed it to discover*.

There are actually no grounds for the absolute worship of causality, the governing principle of the external universe, and certainly none for automatically imagining it to hold sole sway over human lives. For if, as I suggest, the laws of the universe are, in one sense or another, the laws of the Ego, and if it is further true that the laws of the Self reverse all such laws, then we may look for – indeed my view demands – a second and opposite governing principle.[6] This we appear to have in C. G. Jung's a-causal principle of *synchronicity*.[7]

I am unable to give here any neat definition of what I consider synchronicity to involve, largely because my thinking on the subject

---

[6] Logically this should of course not be a governing, but a governed, principle – and that proves to be the case.

[7] I suggest that Freud's Pleasure Principle is a special case or a sub-case of Jung's broader formulation.

is unfinished. Neither do my views in the matter wholly agree with those of Jung, or of others. Following Jung, however (and in terms now least affronting to the Ego), I suggest synchronicity may be considered to have to do with the total configuration of events in time and space which make up a 'moment', and with certain attendant properties of the moment as defined. At any one point in time an unspecifiable number of events momentarily coexist – and not for any causal reason in the normal sense of that term, though certainly each of these events itself has a causal history. As Gestalt psychology has shown (and I am of the opinion that much of what preoccupies the Gestalt psychologist is relevant to synchronicity), a particular whole – a pattern or whatever – may confer on a part values which it does not possess in isolation. (This, as we see, is then a governed position.) Apart from their objective and more or less stable values, then, real – that is to say, non-fantasy – events may possess a virtual infinity of values which are dependent or conditional on transitory, and in some cases never, or at least rarely, repeated, situations. Philosophers point out that a man can never step into the same stream twice. But actually, of course, not even does the same man step into the same stream twice.

Part of what I propose is that at some unconscious level a knowledge of the values of *some* of the events making up a 'moment' may, variously, enable us (a) to envisage the total configuration, (b) to understand the 'conditional' value of other known events in the total configuration, or, (c) still more importantly, to understand the nature and value of events which we do not in any direct sense even know to exist – but which must necessarily exist, and so exist, if the known events are to be as they are.

The foregoing could do with very considerable amplification – but in fact I do not wish to open up these particular aspects of our subject either in this summary or in the present book. One sees already, however, the relevance of the concept of the 'moment' and of synchronicity to such activities as card-reading, fortune-telling and astrology. Again, with that, I do not wish necessarily to endorse these activities. But certainly I believe that they represent attempts to approach or appreciate the phenomena under discussion. The Chinese *Book of Changes* (the *I Ching*) represents a much more sophisticated and, as I believe, at least partly successful attempt to the same end. Following Jung, I am also of the opinion that the general phenomena of *extra-sensory perception* too are best understood within this same general framework, and emphatically not within the rubric of causality. The failure of orthodox science to deal with these phenomena would then be rather inevitable, since its very premises are inappropriate.

The suggestion, then, is that the principle of synchronicity exercises the same overall predominance in the affairs of the Self (the unconscious) as causality exercises in the affairs of the Ego – in waking consciousness and the material world. Here again, the notion of synchronicity purely as a psychic or subjective principle, operating only in the inner existential world of man, might not perhaps prove too unpalatable an idea to the natural scientist. The question – and the real problem – however, is whether the principle of synchronicity in any sense operates *also in the actual physical universe*. It is attempts to argue this last position which especially tend to outrage the scientific establishment.

Further examination and discussion of the particular natures of the Self and the Ego, however, and the nature of phenomena in general in the light of such discussion, enable us to see that what appears to be a head-on and irretrievable clash can be at least partly side-stepped – to see that the clash *itself* is not a real or objective problem, but simply a piece of human behaviour, arising out of the functions of our own personalities – an impasse created purely by terms of reference.[8]

First, however, let us attempt to throw out the whole proposition, by asking what business one has in attempting *at all* to introduce the subjective affairs of the Self into the objective affairs of the Ego. For if it is true, as claimed, that the physical, external universe determines, and has determined, the nature of the Ego, and if the Self always reverses the attributes of that persona, it must follow that the *Self* will determine the nature of the *universe*. Far from understanding the universe, the Self will merely project its own laws (or imaginings) on to the real world. In brief, this is precisely what does happen. So when the Self looks at a tree, a mountain, or the moon and sun, it considers them to be animated by a spirit or consciousness like its own. And believes itself furthermore to be centrally important in the universe – that the universe is actually somehow designed for the benefit of man. Or, even when there are felt to be gods or spirits in charge of the universe, who are in a sense more than man, these are nonetheless held to be preoccupied and involved, benevolently or malevolently, with human affairs.

It is not necessary to attempt here any defence of such attitudes – though in *Total Man* I was at some pains to show that Self pronouncements of this kind are not nonsensical, but on the contrary have a meaning, a value and a genuine informational content. Neither shall I here argue, along with Freud and others (beyond simply drawing attention to the point), that when a person believes something to be so then it *is* in a very real sense – Freud's 'psychical reality' – even if false on the objective view. Nor again ask, legiti-

---

[8] The position has been well described by Friedrich Schiller in his book *On the Aesthetic Education of Man* (1793), and I shall be discussing that work in some detail in this present volume.

mately, with quite what justification anyone should take aspects of the material, objective world to be more real than aspects of his own existence. Instead, we shall turn to a consideration of the conditions under which the cerebellum and sleeping consciousness evolved – or rather, began to evolve. For while, as C. T. Morgan and E. Stellar have put it, 'it looks as if nature started out to make the cerebellum the highest centre of the nervous system', it seems on the other hand that she 'changed her mind and developed the cerebrum instead.'[9]

Although I maintained earlier, and shall continue to maintain, that the Self projects its own laws and characteristics on to the universe, it is nonetheless clear that neither the cerebellum nor its psychological partner the Self evolved in any kind of vacuum. Like the rest of the brain (and the Ego) they were quite definitely in the physical universe and a part of it. The external universe did, then, and in fact must have, played some part in the development of the Self (just as the Ego on its side makes some contribution to the 'laws of the universe' and to our understanding of them). Indeed, it would appear that the cerebellum (the developing Self) originally began the same process of turning out towards the physical universe as did later, and with such signal success, the cerebrum and the emerging Ego. The evolution, in association with the mid-brain and the cerebellum, of an additional pair of pineal eyes placed on top of the head is, as already stated, thought to have been fully carried through by some extinct reptiles. (Today, the much atrophied pineal gland is actually still light-sensitive in the adult lamprey and in the larvae of some amphibians.) While nature's attention – particularly in respect of the coming-to-terms with the external universe – then shifted, as stated, to the developing cerebrum, the cerebellum did nevertheless continue to evolve on its own account. Here is a paradox. For this now apparently neglected and thenceforth largely superfluous organ continued to evolve, *reaching its highest level of development in man himself.*

Whatever else, the depths and even the shallows of the sea, where our distant ancestors and life itself came into being, are not well suited to the development of visual acuity. In the depths there is little or no light, at the surface there are problems of refraction, and so on. Instead, sound and smell are under water the main methods for acquiring information at a distance. This holds true even for those mammals which have returned to the sea – whales, dolphins and so forth (who also possess, as it happens, especially well-developed cerebella). These are distinguished by extremely fine powers of hearing. The shark, on the other hand, is noted for

---

[9] C. T. Morgan and E. Stellar, *Physiological Psychology*, 2nd edition (McGraw Hill, New York, 1950), p. 288.

his astonishing powers of smell. Indeed, the first part of the cerebral cortex to emerge to prominence – in the fish – is the olfactory lobe. May one not ask, however, whether at that time the *cerebellum*, perhaps with the co-operation of the mid-brain, began to evolve still other powers of 'perception at a distance' – powers which were to be abandoned, half-developed, along with the still larger plan for a cerebellar HQ.? And might these powers perhaps have had something to do with synchronicity in the first place, and in the last with extra-sensory perception?

With the subsequent emergence of our fishy ancestors from the sea, a realm is invaded where light, not odour or sound, is far and away the best means by which to investigate and map the environment, with direct vision as receptor. Yet though light is the best of the available possibilities, its election was not wholly inevitable. Dogs, for example, have only poor powers of vision and rely heavily on smell, while other species like bats, virtually blind, make extremely effective use of sound. (But one can neither smell nor hear a star.) In the ancestral line which produced modern man it was vision which came more and more to the fore, and indeed at the expense of the remaining senses.

What, then, of the cerebellum? As noted, its evolution, counter perhaps to reasonable expectation, continues. But if its original direct links with the external world were limited, even this tentative contact now ceases entirely. Darkness closes in upon the cerebellum. Such destiny as remains to it, it must work out, buried beneath the cerebral cortex, sightless and alone. And yet matters are not quite so desperate as they seem: for the Self *learns to make use at second hand* of the information which is directly garnered by the Ego. This is an extremely important circumstance.

Viewing the personality, we find the Ego, like the pilot of an aeroplane, looking directly out on to the external universe from a high, frontal position, while the Self lurks behind and within. To this circumstance, I suggest, such Self figures in legend as the Follower and the Doppelgänger owe their existence, and still others (rather less nebulous) such as Lot's wife and Eurydice – for whom one may not look back.[10]

The Self, as indicated, is blind, having only vestigial eyes of its own. Out of this circumstance arises, I believe, the great psychological power of such figures as Blind Pew in Stevenson's *Treasure Island*, and our general fear of all secret and underground creatures – at the same time, too, our empathy with figures like Samson, 'eyeless in Gaza'.

---

[10] Satan, too, is told 'get thee behind me' – that is, back to his proper place.

The Self, then, peers out at the universe as it were *through* the Ego. 'Peers' may seem a strange choice of verb in view of what we have just said. Yet, as stated, the Self does receive independent sensory messages from many parts of the nervous system, which may in some sense or other still be experienced visually (as, by way of parallel, simple pressure on the closed eyeball is experienced as colour). Secondly, however, there is good reason to suspect that the Self has direct access to the visual memory store of the cerebral cortex, of which, it seems, full use is made during dreaming. In either or both of these ways the Self could be said to see the outside universe. In these general circumstances we have one of the meanings of the phrase 'second sight'.

Having thus far emphasized the subsidiary or secondary position of the Self in the psychological economy of the personality, we must now redress the balance somewhat. For the Self and the Self-system do (or certainly can) exercise a strong influence on waking consciousness, in the case of some individuals actually a dominant influence. Before going on to discuss this point one or two other matters must be raised.

The two psychological–physiological systems – Systems B and A – which respectively underpin the Self and the Ego, can also function in the absence of their appropriate consciousness. They are then said to function *robotically*. This term is not to suggest, however, that they then function in any way unsubtly. The work of monitoring, coding, collating, and so on, by each of the systems, then, continues, even while the opposing consciousness is in overt control.[11]

A further point is that both Systems – perhaps particularly System B – are in some ways 'empowered' (pre-programmed) to disrupt the conscious activities or intentions of the legitimate regent when circumstances sufficiently warrant. The 'robotic' interference of System B may be seen most clearly in the behaviour of neurotics. However, in normal individuals, at one end of the spectrum, one can cite the onset of overwhelming panic which sweeps away our conscious intention to stand our ground or to act calmly; at the other, the well-known Freudian slip, where one says something other than one intends to say, or believes oneself to have said. The robotic interference of System A in System B affairs is demonstrated when one wakes up during a frightening dream or when the name of a friend is called out (even though other noises of the same intensity

---

[11] Perhaps the point should once more be made that the term 'consciousness' in this book in general refers equally to waking consciousness and to what is normally termed the unconscious. The latter is properly referred to as 'night' or 'sleeping' consciousness – though I myself also continue to employ the term 'unconscious' for reasons of general convenience.

and the names of strangers fail to produce that effect); and perhaps most dramatically in sleep-walking.

We are here primarily concerned, however, with the 'interference' by System B during waking hours, an interference that is so pervasive, and on such a scale, as to have the effect of imparting a particular quality to entire sections of some individuals' thought, behaviour and attitudes. The person habitually so affected I term a *System-B-dominant*. All continuous, consistent behaviours of the B variety (as opposed to the occasional Freudian slip, and so on) I refer to as 'trance conditions', and these take many forms. Nevertheless, despite a wide diversity in type and form, all these B-behaviours, or trance conditions, exhibit overriding similarities and numerous points of correspondence. Were this not so, of course, one would have no justification for placing these behaviours together in one general category.

A clear instance of a trance condition is the state of hypnosis. The entranced subject is prone to obey the commands of the hypnotist and has largely lost his own conscious, independent volition. That the unconscious, and not waking consciousness, is involved here is clear from numerous circumstances – for example, in the fact that the hypnotized person usually recalls nothing of what has transpired on reviving.

A second fairly obvious instance of a trance condition is sleep itself, and in particular the state of dreaming.[12] Although we speak of having a dream, it would be more accurate to say that the dream has us. For, during dreaming, in contrast to the (usually) consciously directed actions of our waking life, it is we who are under the direction of the dream. In general, we must follow where it leads and observe what we are shown.

In applying the term trance condition to the states of hypnosis, sleep and dreaming one has not exceeded the normal values of the word 'trance', nor indeed proposed anything of a particularly revolutionary nature. It is with the instances which follow that new ground is broken, and the word *trance,* while not, however, losing its more normal connotations, is raised to a technical term. Other trance conditions, then, are : religion; neurosis; socialism; communism; being in love; being a woman; and being a child.

The full discussion of these matters occupies a considerable part of *Total Man* and can hardly be repeated here. The briefest recapitulation of the main points, however, is as follows.

Neurosis resembles both dreaming and hypnosis in that volitional

---

[12] If we may already begin to indicate connections between the various trance states, the literature generally agrees that women hypnotize more easily, and sleep and dream more than men.

consciousness is significantly overridden. The yielding-up of volition and other aspects of individual assertion is a feature also of most religions and of communism.[13] Women and children, it appears, suffer rather more from neurosis than do, respectively, men and adults. More women than men are interested in religion, particularly in what we may call ecstatic (or hysterical) religion – more women than men are spiritualist mediums, for example. The Socialist Party of this country, again, has, and has always had, a greater number both of female parliamentary candidates and of elected female M.P.s than the Conservative Party. A further common feature of socialism and communism is their treatment of women as equals. In general, young people seem more strongly drawn than adults to Socialism (with adults probably more Conservative). A further System B product not so far mentioned, itself a trance condition, is psychoanalysis. Religion, communism and psychoanalysis all involve belief rather than proof. Each, for example, features a 'sacred' body of literature which is handed down and, though annotated and extended, never basically questioned or refuted. (One thinks of the Bible, the Jewish *Torah*, the Hindu Vedas, the writings of Marx, Lenin, Freud, Jung, and so on.) All three of these activities speak in their differing ways of a golden age or millennium to come: religion in terms of the reappearance of God or of the rejoining of the world with God; communism in terms of the fall of capitalism and the still more ultimate withering away of the State; psychoanalysis of the rebirth and reintegration of the individual in his own lifetime.[14]

As has been seen, not all System B or trance conditions exhibit *all* the suggested common features – or certainly not to the same degree. We must think not so much of an all-or-nothing situation but rather of an interweaving or interlocking of certain tendencies, involving greater or lesser overlap. However, one further underlying conviction or sentiment which, I believe, fairly unites *all* the diverse behaviours concerned is that of 'wishing (wanting) will make it so'. I shall not take time here to expand fully on this notion, but one

---

[13] The initial resignation of volition in many of these examples is, of course, itself sometimes, in some sense, a volitional act. One, so to speak, chooses to give up one's freedom of choice. In many cases, however, it is a question more of a seduction – this rather sexual term being used quite advisedly.

[14] I am of the opinion that the psychological origins of all beliefs in a coming millennium or golden age are rooted ultimately in the female urge to give birth to children. Many of the common features of System B, and indeed of System A, behaviours are evolved and modified expressions of similarly basic biological urges. I shall be developing this general idea further in the course of the book.

may readily see its influence (variously) in the Freudian dream-as-wish-fulfilment, the views of Socialists on education (sows' ears can become silk purses), the conviction of the religious zealot that he will reach God (prayer itself is an excellent example of wishing), the wholly explicit granting and fulfilling of wishes in fairy-tales, the lover's belief that love conquers all, the neurotic's wish to be free of his symptoms without usually any willingness to take practical steps towards their removal, and so on.

The broad spectrum of activities defined as System B behaviours is mirrored by a spectrum of System A behaviours. Some of these, in no particular order, are : the sciences and the scientific method; logic; experimental psychology; the practice of magic; fascism; capitalism; business practice; warfare; and being a man.

There is this difference, however – that *two* sub-types of System A behaviour can be distinguished : those which (more or less cynically) manipulate, exploit and dominate System B states; and those which are concerned either with the outright denial or with the *destruction* of System B. These two types of activity are defined respectively as *trance manipulation* and *trance denial*. These two notions help us to understand some of the apparent contradictions in the items of the list just given. In general evolutionary terms, trance manipulation (e.g. magic) is older than trance denial (e.g. science).

The most readily identifiable unifying component in all System A behaviours is their high degree of ego-involvement. Each in its particular way evidences the triumph of the individual over odds and circumstances – ranging from the physical circumstances of battle and attack by enemies to the metaphysical challenge of problem-solving in the face of an 'unwilling' universe. Such abstract or metaphysical ego-activity originates (as in the case of the Self) in much more directly biological activities. I share, for example, Desmond Morris's opinion that gambling is a form of hunting,[15] and would extend the view to include aspects of business and business practice, the profit, for instance, representing the kill. I am further indebted to Charles Reich in this connection for pointing out how rich the language of the boardroom and the stock market is in battle and imperial terms – so one speaks of oil barons, tobacco kings, a killing on the Exchange, the Woolworth empire, market strategy, and so on.[16]

Two extremely important attributes of the Self and the Ego are the *archetype* and the *archestructure*. The former is C. G. Jung's term for certain universal, genetically inherited response-tendencies

---

[15] See *The Naked Ape* (Jonathan Cape, London, 1967).
[16] See *The Greening of America* (Allen Lane, London, 1971).

to certain configurations and circumstances, especially in the context of inter-personal relationships, common to all human beings.[17] Without essentially departing from Jung's conception, I myself use the term archetype to refer to the *psychological correlate* – the conscious mental experience or image – associated with and somehow arising from fixed response patterns which exist at the physiological level. I believe that archetypes in human beings are in many ways the equivalent of the 'sign-releaser' in animals, in the presence or absence of which animals respond or do not respond in particular ways.[18]

The archestructure is a concept of my own. It is not unrelated to the notion of the archetype, and hence deliberately echoes Jung's earlier term. I define an archestructure as a somehow perceived or felt attribute or function of the nervous system, acted out or 'discovered' (actually, of course, in some sense of that term, projected) in the physical, social and cultural environments.

Instances of archestructures include the already mentioned cross – both the Christian variety and the much older swastika, as well as the widespread symbolism of the cross-roads, crossed knives, cross-bones, and so on – these all being, essentially, as I suggest, representations of the crossing-over of impulses between cerebrum and cerebellum. The political organization of Western democracies is also largely archestructural in origin. The opposition of the political Left and Right is, whatever else, an expression of the inner psychological tension between Self and Ego, between System B and System A in each of us.[19] As a final instance, I propose that the moon and sun in legend, mysticism and religion, and their attributes in those contexts, are archestructural. The moon is the cerebellum, the sun the cerebrum. The qualities of the sun – brightness, its intimate connections with waking consciousness, vision and the day world of reality, the fact that it rises and sets when it does (i.e. when the cerebrum rouses and sleeps) – are also the qualities of the cerebral cortex *as perceived by the Self*.

With this last statement we arrive at one of the not wholly resolved aspects of the general theory. In *Total Man* the position

---

[17] Jung says rather that the response tendencies *produce* these configurations – and, of course, I would agree that there is a two-way traffic involved here. (See Jung's *Collected Works*, Vol. 9).

[18] See, for example, N. Tinbergen's *The Study of Instinct* (O.U.P., London, 1951) and Konrad Lorenz's *King Solomon's Ring* (Methuen, London, 1952).

[19] It is of interest that the political party not in power is called the Opposition – this word of course deriving from 'opposite'. The choice of term here is again itself unconsciously governed. The designation of Socialism as 'Left' is similarly so determined, a reference, I suggest, to the 'left-handedness' of the cerebellum.

more or less adopted was that the perception or experience of archestructures was solely an attribute of the Self, and the archetype a property of the Ego. The suggestion was that only the Self could perceive the nervous system as a whole (by virtue perhaps of its inner position, or the fact that it is, in a sense, the original occupant of the house now principally occupied by the Ego) – and, in addition, itself. Archetypes (that is, sign-releasers) on the other hand are what the biologist calls species-specific.[20] They are acquired in the course of the very process which leads to the emergence of a new species from some earlier evolutionary stage. It is likely, then, in fact obligatory, that the physiological infrastructures concerned will be located in the most recently evolved parts of the nervous system – that is, in the cerebral cortex. This, I suggested, would tend to link the physiological and psychological structures in question very closely with System A and the Ego.

Without altogether revoking the foregoing I would like nevertheless to amend it. I am now persuaded, on the one hand, that the Ego also produces or experiences *archestructures*. The further nature of these I shall be discussing later. On the other hand, as I already tentatively proposed in the earlier book, I now firmly believe that the Self for its part produces or possesses archetypes. The *cerebellum* has certainly added to parts of its own cortex in relatively recent evolutionary time (in parallel with the cerebrum) so that a possible locus for the now suggested archetypes exists. What give at least some initial support to this notion are, for example, experiments which show that, in at least some species, female animals with the cerebral cortex ablated can perform general sex (and certain specifically parental) acts without much loss of fineness, while male animals so treated cannot.[21] The physiological basis or infrastructure of these behaviours must then necessarily be located, in the female, elsewhere than in the cerebral cortex.

Most human archetypes are likely, then, to have been acquired in the course of our recent evolutionary history. They are responses, or more precisely, the psychological accompaniments of responses, which have become genetically fixed – principally, I suggest, because those who made such responses tended to survive better than those who did not, along normal Darwinian lines.

The period of many hundreds of thousands of years during which *homo sapiens* emerged from the ranks of the primates is therefore the one of greatest interest to us when considering the appearance

---

[20] Of course, in the strict sense we can never *know* whether animals have any psychological or mental equivalent of the archetype – even supposing that sign-releasers and archetypes are the same phenomenon, as I suggest.

[21] See C. T. Morgan and E. Stellar, op. cit., Chapter XX.

of inter-personal and social archetypes in the human mental landscape. When we examine the period in question we find an extremely interesting situation. For it appears that during that time human history consists of two separate, though inerwoven, strands.[22] It seems, in my terms, that nature produced both a B-dominant and an A-dominant variety of sapient.

A variety, one must emphasize, is not yet a species, though it is on the way to becoming such. Separate varieties, for example, are still capable of cross-breeding with each other, while species in general are not. My contention is, in the case of man, that during long periods of relative isolation, in certain parts of the world, variety-specific archetypes, and other variety-specific behaviours, evolved in, principally, two kinds of sapient.

In what is termed *paleoanthropic* or 'ancient' man we have, as I shall describe, a relatively pure strain of B-dominance, while in *neanthropic* or 'new' man we have an A-dominant strain. These two varieties are contemporaneous on the planet over a considerable period of time. The notation of B-dominant and A-dominant is, to repeat, purely my own. The terms paleoanthropic and neanthropic are not, nor are the well agreed physical characteristics on which these specialist classifications are based.

The Neanderthaloids of Europe and the Near and Middle East of some 50,000 years ago are one instance of a paleoanthropic variety. The general characteristics of this variety are, however, seen at their most exaggerated in the so-called classic Neanderthals of western Europe. The classic Neanderthal was short (average male height around 5 feet 4 inches), barrel-bodied and squat, with extremely powerful though curved arms and thighs. His forehead retreated sharply from prominent brow-ridges, while the rear of his head was curiously enlarged and extended. I have proposed that this rear development housed an enlarged cerebellum, as well as larger temporal and occipital lobes. The temporal lobes in modern man are concerned with memory, the occipital possibly with imagination – since, for example, this part of the head emits an altered brain-rhythm when the eyes are closed and day-dreaming is in progress. I have also suggested that the Neanderthaloids, and classic Neanderthal in particular, were nocturnal or semi-nocturnal in habit. The

---

[22] Once again, I cannot repeat here the lengthy discussion of this matter undertaken in my earlier book. It must suffice to say that the view I hold has the support of a number of respected zoologists, anthropologists and archaeologists, including A. S. Romer and L. S. B. Leakey. It is, however, not the only view possible nor held in a field of study where in any case we have by no means as much information as we should like.

large, round eye-sockets, among other items, do suggest eyes used to making the most of reduced illumination. The curved arm and thigh bones together with various other anatomical peculiarities again argue endemic adult rickets in a people seldom exposed to sunlight, one of the few natural sources of Vitamin D.

The neanthropics of approximately the same period are represented by Cro-Magnon, a sapient in many ways resembling our own physical type. He was tall (average male height around 6 feet), muscular but athletic and to us of aesthetically pleasing proportions. The skull showed a high forehead with virtually no brow-ridges, and a flattened rear or occiput – in contrast to the enlarged occiput of Neanderthal. The frontal lobes of the brain are considered, in modern man, to be the seat of the higher cognitive and logical processes. It is these which were apparently less well developed in the Neanderthaloids, and relatively more developed in Cro-Magnon.

Some – though always progressively fewer – authorities believe that Cro-Magnon evolved rapidly from a Neanderthaloid stock over a very short period. The majority proposes that Cro-Magnon invaded Europe from some as yet unknown locale, the standpoint which I myself adopt. The only serious argument against the view is the fact that the (alleged) home-site of Cro-Magnon has so far escaped discovery.

At all events, it is agreed by the specialist literature that Cro-Magnon reached western Europe some 35,000 years ago in the wake of the temporarily retreating glaciers of the Würm ice age. There in probability he made the acquaintance of classic Neanderthal – a Neanderthaloid cut off from the main body of humanity for some 40,000 years by extensive glaciation. In the conditions of extreme cold and privation certain physical and behavioural characteristics had apparently proved favourable to survival – for example, the short, thick body – and these had been reinforced by normal processes of natural selection. Coincident with the arrival of Cro-Magnon in western Europe, classic Neanderthal abruptly disappears; while we have, and probably will only ever have, little direct evidence of this, it is very possible that the latter was slaughtered by the former. This, in any case, is the view which I myself adopt.

I have suggested that, in classic Neanderthal, Cro-Magnon was confronted by the embodiment of his own 'animal' nature and origins – the Self, in fact. I have suggested that this confrontation was deeply shocking and repugnant to him from many points of view. For instance, it is probable that Neanderthal was (ritually) cannibalistic, and quite certain that he had evolved a form of what we today would call magic, perhaps 'black magic'. With classic Neanderthal we find for the first time such practices as ritual burial of the dead – arguing for the belief in a detachable soul – and so on.[23]

One envisages a period – perhaps extending over several thousand years – of systematic slaughter of Neanderthal by Cro-Magnon. No doubt Neanderthal resisted. But we may assume a greater degree of organization and strategic planning on the part of Cro-Magnon (together perhaps with a fierce love of battle) that the other lacked. During this relatively long period I suggest that survival favoured those who fought best – those, too, who were most affronted and fanatically opposed to Neanderthal: in short, that an archetypal (and antagonistic) response to the essential attributes of classic Neanderthal, and therefore in milder form to those of the Neanderthaloids generally (and so, too, in some sense to attributes of the Self), became biologically fixed in these, our own immediate ancestors.

I have tried to show that the stories of dwarves and trolls in both myth and legend, and for that matter of the witch herself, are 'accounts' by Cro-Magnon of his encounter with classic Neanderthal.[24] The parallels are too numerous to detail here. But if we briefly imagine a strange and (from our point of view) physically repellent people, given to moving about at night and hiding by day, latterly found only in the most inaccessible parts where they have gone to avoid slaughter, in valleys high up in the mountains, or deep in caves, or in impenetrable forest, the men stunted, though hugely powerful, with curved limbs and hunched backs (a probable consequence of adult rickets) – do we not see already more 'coincidences' between myth and reality than we can readily account for on a chance basis? I believe, in particular, that the very prominent brow-ridges of classic Neanderthal are the original model for the horns of the Christian Devil.

If this view of the myths and legends in question is correct, can they have survived some 35,000 years of oral transmission – for, of

---

[23] Or to be more precise, in the light of Professor Solecki's recent discoveries, for the second time. Cf. *Shanidar: The Humanity of Neanderthal Man* (Allen Lane, London, 1972).

[24] Conversely, I have proposed that parts of the Bible, the Apocrypha and the *Pseudepigrapha* are a stylized account by Neanderthaloids of the passage of Cro-Magnon through the Middle and Near East some 35,000 years ago. The former, I suggest, are the 'giants' and 'fallen angels' of those texts. Yet other parts of the Bible are not 'mere', and since that time distorted, chronicles of actual events, but are, I believe, intuitive and codified statements concerning evolutionary and biological matters. For example, I suggest the story of Cain and Abel to be an analysis (rather than an account) of the clash between the Cro-Magnon and Neanderthaloid types (and, as ever, between the Ego and the Self).

course, writing did not exist? I myself believe that they probably would not have survived without the archetypal and archestructural support I have proposed. Only because these stories awake inherited or built-in echoes from the depths of our psyche do we listen to them, in a sense understand them, and repeat them to our children.

Eventually, the pure Cro-Magnon variety of man who occupied western Europe after the demise of classic Neanderthal was himself slowly overrun and infiltrated by varieties (or sub-varieties) of men having varying amounts of Neanderthaloid blood in their veins. These were, as I see it, the descendants of the half-breed children which I propose Cro-Magnon left in his wake, along with whatever of his number had 'gone native'. These diluted Cro-Magnons were probably more or less acceptable to Cro-Magnon proper – that is, did not as a rule trigger off the violently adverse archetypal reactions; though, I suggest, these individuals would nevertheless in general have been regarded as 'inferior'. At any rate, only at the fringes of Europe, furthest removed from the Neanderthaloids and the later westward-moving part-Neanderthaloids, do we find today pockets of individuals recognizably Cro-Magnon in physique – in Scandinavia, the Canary Islands and elsewhere.

I believe that we, modern man, are a hybrid, and indeed an extremely unstable one – the product of the crossing of a relatively pure A-dominant (Cro-Magnon) with a relatively pure B-dominant (the Neanderthaloids), though not, incidentally, in equal parts. For the most part, modern man is principally and recognizably neanthropic. Not only is the mixture by no means even, however, but the precise quantities of the components vary also from one part of the world to another.

Briefly, as just stated, I recognize modern man as essentially neanthropic. However, relatively speaking, the Negroid and Asiatic varieties are today nearest the paleoanthropic in type, or, put another way, show the greatest paleoanthropic admixture. The Asiatic, I suggest, is (and always relatively speaking) more properly what I term a B1-dominant, while the Negro is a B2-dominant.[25] (See below for an explanation of this distinction.)

I have taken it to be significant, however, that it is in the areas where the mix was originally most even – the Near and parts of the Middle East – that what we term modern civilization later emerged and flourished (farming and permanent settlements, ultimately the

---

[25] From Henri Breuil and Raymond Lautier's *The Men of the Old Stone Age* (Harrap, London, 1965). I learn that in modern Asiatics the cerebellum is 'very incompletely covered by the brain' – an astonishing confirmation, I feel, of my view of the greater *metaphorical* prominence of their cerebellum.

first city, Jericho, the discovery of metal, hand-writing, the great world religions, and so on) and *not* in western Europe, the home of the pure Cro-Magnon strain. In saying this I do not, incidentally, dispute that either variety alone could and perhaps would in due time have evolved some kind of relatively advanced civilization. This would, I suggest, nevertheless have been in both cases a very one-sided and predictable affair. What we term the man of genius and the true artist would, I think, have been lacking. For my own view is that it is the mixture or hybridization suggested which produces for example Faust's 'two spirits within one breast' – the creative conflict which is at once our blessing and our curse.

At this point some further elaboration of the general theory may be introduced. While Systems A and B remain the two major personality divisions, there exist also some important sub-divisions. System B is further divided into Systems B1 and B2. These once again correspond to – or have evolved in association with – two identifiable physiological sub-systems. It will be recalled that System B as a whole was said to be the correlate of the autonomic nervous system. Systems B1 and B2 are in similar correspondence with the sub-divisions of the autonomic system – the para-sympathetic and the sympathetic nervous systems.

The roles of these two physiological sub-systems in the human economy are briefly as follows. The para-sympathetic conserves and stores bodily resources. It encourages the flow of gastric juices and saliva – that is, aids digestion – facilitates urination, defecation and aspects of sexual readiness, is concerned with body-temperature and sleep and, importantly, contains what are termed 'pleasure centres'.

The sympathetic system, in contrast, spends resources. It mobilizes the organism for the emergency reactions of fight and flight. So it augments the blood supply to the muscles and brain at the expense of the stomach, speeds up the heart-rate, and temporarily inhibits the vegetative functions of digestion and similar processes.[26]

From this brief glance at physiological function many of the broad *psychological* attributes of Systems B1 and B2 emerge. It is, however, not my intention to link my statements about the psychological sub-systems in every case to their physiological sub-stratum,

---

[26] It should be emphasized that these two physiological sub-systems always function in close conjunction and only rarely in isolation from each other. There is much less of a split here than is found between the total autonomic and the central nervous system. I have suggested that this may be one of the reasons why at the psychological level single emotions tend to be hard to isolate, and one readily accompanies, turns into, or is replaced by another.

The Background

even if that were possible. Instead, at this point, we entirely revert to the psychological level.

From one point of view I consider that Systems B1 and B2 together comprise the unconscious, and that System A by itself contains waking consciousness. From another standpoint, however, there are good reasons for considering that Systems B2 and A taken together make up the Ego or the 'male organism':[27] while B1 alone constitutes the Self and the 'female organism'. The total cake, then, can be cut in two ways. Or, to express it slightly differently, System B2 has something of a double role in the economy of the personality. Figure 4 shows this two-way division schematically.

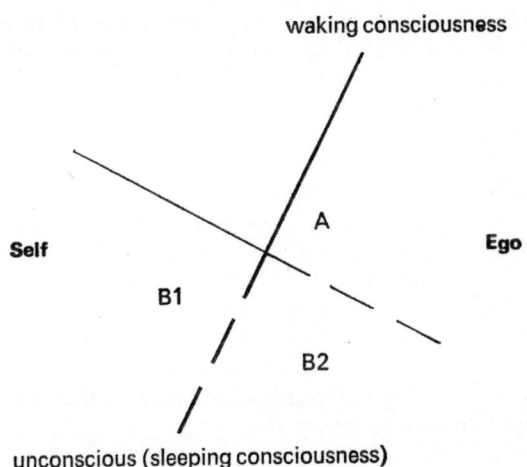

FIGURE 4
Schematic representation of the relationship of Self, Ego, the unconscious and waking consciousness.

I am of the opinion that System A evolved directly from System B2.[28] This view receives support from theories of body-types, to be discussed. If this is the case we would look for many literal con-

---

[27] Freud himself, it should be mentioned, considered the ego-instincts to be located within the unconscious, where, nonetheless, he separated them from the libido-instincts. My own model therefore parallels Freud's on this point.

[28] One can fairly readily accept that some thinking at least – i.e. mental problem-solving – has evolved from actual physical movements. This kind of thought involves *actions undertaken symbolically* or notionally, instead of literally and actually.

nections between those two systems, such as emerge in the following considerations.

The mode of the Self is Being, while the mode of the Ego is Doing. On this view we see at once that B2 belongs properly with A – for both flight and fight (B2) involve doing, and these actions take place in the real world. (That fight or flight nevertheless have, or can have, a strong unconscious component may be seen on those occasions when we realize, to our conscious surprise or shock, that we have for example hit someone with whom we were arguing, or are running away from some situation, are screaming, and so on.)

A few words may be said on the subject of Being. So dominated are the West and western thought by the Ego that we have almost lost sight of the concept, let alone the experience, of Being. Nonetheless, it exists as a central idea in many cultures of the East, notably perhaps in India[29] and in the China of earlier days. We have gone so far even as to produce active verbs for many conditions which are really states, as if these *were* actions. Thus, for instance, we speak of 'to love' someone. Yet no one could 'love' if you ordered them to do so. (They might, of course, make love, which is something else again.) Properly we should, and do also, speak of 'being *in* love'. The example I have just given does perhaps leave room for counter-argument. Let us therefore consider a clearer instance. In the Christian religion a person who is at one with God is said to be 'in a state of grace' (cf. to be 'in love'). But this time there is no active verb available. There is no way you can 'do' a state of grace. With these all-too-brief remarks on that subject, however, we move on to other matters.

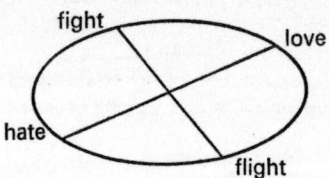

FIGURE 5
The emotional 'compass'.

I proposed in *Total Man* each of the sub-systems B1 and B2 to have a positive and negative pole – one of the ways in which B2 is closer to B1 than to A, since System A proper has almost no such polarity. In the case of B1 these poles are love and hate. With B2

---

[29] One of its most lucid exponents, from a Western viewpoint, is Krishnamurti.

they are fight and flight.[30] It is possible conceptually – and, I think, more than simply conceptually – to arrange these as points on a compass, as in Figure 5.

Partly, as I believe, through the agency of blanket concepts such as aggression, we tend to equate hatred with fighting, and vice versa. Yet is it perfectly feasible to fight and even kill without hatred. We see this in the knightly 'passage of arms'. We see it again in the idea of sportsmanship where 'the best man wins' without rancour from the defeated. Despite modern cynicism, I believe this state of mind to be genuinely achievable.[31] It is also, perhaps less obviously, quite possible to hate without ever *doing* anything or even wanting to do anything about it. Thus a miser may hate those with more money than himself, or the spendthrift. He may not wish the spendthrift any physical or even mental harm.[32] It is the *notion* of thriftlessness which is disturbing and hateful. An example of the reverse emotion, in isolation from action, is possibly the idea of God's love for us as a constant which exists in the universe quite independently of what we do, or however many times we sin. Again we come back to the idea of love and hatred as *states*, not actions which can counter-*act*.

I take the poles of love and hate to be those of the receptive-passive 'female' personality, and fight and flight as those of the active 'male' personality. Of course, both the actual male and female human beings possess both of these sub-systems. It is a question of which is generally more dominant. And of course again, more than one of the various emotions associated with the poles may on occasion be present at one time, or in very rapid succession. Staying with our analogy of the compass, we might say that in times of emotional disturbance the needle oscillates and circles wildly, much as does the needle of a real compass during a magnetic storm.

With these relatively simple models and analogies, incidentally, I do not wish in any way to deny the extreme complexity of the human organism, or in any way to invalidate the sometimes astonishing subtleties and shades of human psychological states. And yet, I suggest that a good number of the more subtle contents of mental

---

[30] Perhaps some ghost of this polarity is seen in the eagerness to tackle especially intellectual problems or, conversely, the easy discouragement in the face of such. But, strictly, in the purest form of mental inquiry one is discovering (or failing to discover) *facts* (cf. the processes of logic) not *fighting* a problem.

[31] The title of Rommel's memoirs of the African campaign of the Second World War was, interestingly, *Krieg Ohne Hass* – 'War Without Hate'.

[32] Cf., perhaps, George Eliot's character, Silas Marner.

life can be seen to derive, at least possibly, from the simple beginnings I describe. Thus, pride and independence, for example, are in a sense extractable from the combat situation, i.e. 'when I fight I win' and 'when I fight I fight alone' – that is, in neither case need any help. Quite other qualities, such as helpfulness and tolerance (that is, non-dismissive tolerance), can be argued to be not so very removed from love – and so on.

For purposes of further illustration, the basic emotional personality may be diagrammatically represented also as shown in Figure 6. Here small and capital letters indicate the greater or lesser influence of the component concerned.

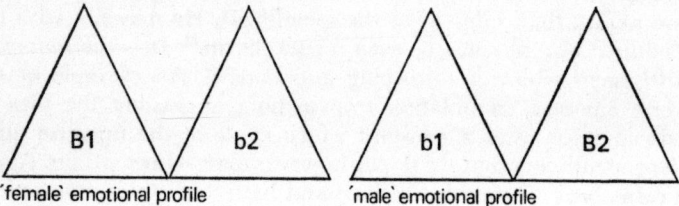

FIGURE 6

If we now incorporate System A – the non-emotional component – we have the situation in Figure 7 :

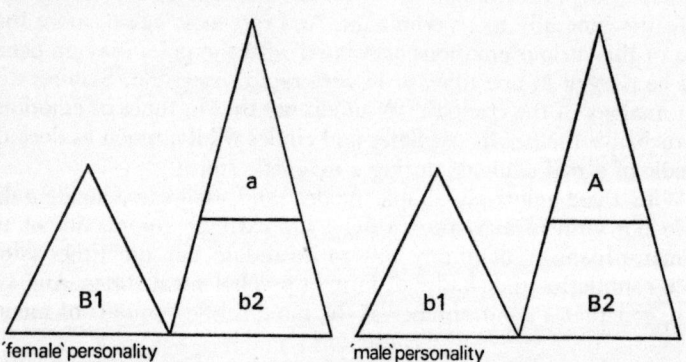

FIGURE 7

The suggestion is that the male is strong on both B2 and A components, the female only on B1. This is essentially the case – though I do not wish to suggest that the B2 and A components have identical values in all males. Cro-Magnon, for example, was, I suggest, actually a B2 dominant. It is today's scholar, 'boffin' or pedant who is the true A-dominant. These two types are shown diagrammatically in Figure 8. I shall suggest later that these correspond to the 'athletic' and 'asthenic' categories of body-typologists.

In the diagrams broad similarities will be observed to both pyramids and cathedrals, and this aspect of my model is deliberately designed to reflect a situation which is not of my making. I believe, that is, that the actual pyramids of Egypt and elsewhere – with their dumpy, broad-based (broad-hipped?) shape, lack of windows and their tunnels running below ground – are an archestructural representation of the cerebellum (perhaps also of the womb and the viscera); while the cathedral – its high tower filled with light-receiving windows – is an archestructure of System A (as perceived, however, by System B). Once again, I cannot here go into all the bases for these assumptions. But if one protests, for example, that the cathedral has nothing to do with System B2, from which I suggest System A evolved, I would point to the warriors laid to rest within it, and the coats of arms which line its walls. How accurately the archestructure here reflects the evolutionary processes involved.

A further point may now be discussed. If System A evolved from B2, does any comparable system evolve from B1, and if not, why not? The question could be put another way. If thinking is a form

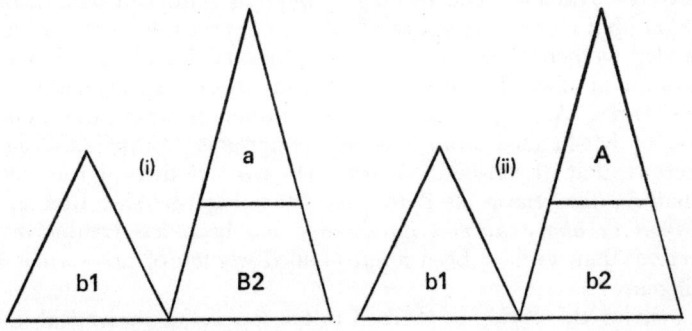

Profiles of (i) Cro-Magnon and (ii) 'scholar'

FIGURE 8

of doing, is dreaming a form of being? And further, if there are 'higher thought processes', should there not also be 'higher dream processes'? The answer is in both cases a qualified yes.

What we theoretically expect to find, but actually do not find to any extent, we may as well call a X-dominant – since we cannot hope to understand too well the nature of the mental landscape involved. The X-dominant, however, is possibly the (unconscious) model for the 'telepath' of many science-fiction and science-fantasy stories.

There are no genuinely B-dominant varieties of man living – the nearest to this type being perhaps the Australian aborigine, or the South African bushman. Yet *because the female of our species* – to emphasize this fact now – is a B-dominant, might we not expect at least a few women to be X-dominants? The precise reasons why so very few, if any, actually are I have argued through in my earlier book. Briefly, I proposed that in the course of the last 50,000 years (and perhaps the last 500,000) men have more or less systematically, though also probably cyclically, killed those women who showed these qualities to any high degree, in one form or another of witch hunt. I suggest that these qualities as a result have been gradually leeched from the gene-pool of modern man.[33] It is, certainly, difficult to prove or disprove such a claim. On the other hand, the witch hunts of the last few hundred years, extensive both in time and geographically, *are* on record; and in the course of those at least far more women were slaughtered than men.

Further, I assume on theoretical grounds that B-dominance and left-handedness are related, and there is some suggestion that left-handedness is more common in primitive, especially non-European (i.e. B-dominant) populations – though reliable figures are unfortunately not available. Yet, turning to modern European populations, we find not merely less left-handedness, but that the number of left-handed women of normal and above normal intelligence is lower than the number of such men – when my theory expects more sinistral women. Among the mentally sub-normal, however, the proportion of left-handed women equals, and some researchers suggest exceeds, that of left-handed men. Does it not then at least look, whatever the reason, as if to have been a left-handed woman of *normal or above normal intelligence* has been less conducive to survival than to have been a left-handed woman of *sub-normal* intelligence?[34]

Leaving this topic, we proceed to the last major theoretical item.

Examining human behaviours and the structures of society, one seems to find, as I have briefly shown, much that can apparently

---

[33] Alternative, if minor, methods which have the same effect as far as the gene-pool is concerned include making such women into priestesses or oracles and forbidding them children.

[34] The question of left-handedness is another that I cannot really develop here in summary. There are some indications, as suggested, that left-handedness in both men and women was once commoner than it is today. And it is true also that in every European language the word 'left' (from Anglo-Saxon *lyft* = weak, womanish) also means evil, unclean, weak, undesirable, untrustworthy, and so on. (Compare, again, the English word 'sinister' – Latin *sinister* = left.)

be accounted for in terms of the attributes and relative dominance of Systems A and B – the Ego and the Self. There are, however, other extremely important behaviours – the most typically sapient behaviours in fact – which cannot at all so readily be accounted for in those terms. These behaviours include art, morality, justice and humour.

Art is neither pure intellect nor pure feeling. It is neither simply form nor simply content. Instead it is a marriage or synthesis of these pairs – and a synthesis, indeed, comprising more than the sum of its parts. The respective contributions of the two sides, one should stress, need not at all be equal. Form preponderates, for example, in classical, content in romantic, art. But both are art.

Justice – as opposed to legality – is not merely the letter of the law. It is the letter (form) tempered or illuminated with compassion or understanding. In the symbol of Justice as a figure holding a pair of scales, or balance (a symbol found also in ancient Egypt, classical Greece and Rome), we see a representation of two parts acting together to produce a third – here a just outcome.

Amplifying briefly, when we speak of a normal individual we often say that he is well-balanced. Justice, as we see, carries an actual balance in her hand. We may also say that he is well-integrated. The Latin word *integer* means undiminished, unimpaired, whole, complete – and impaired means unequal. This whole metaphor and symbolism of the cancelling-out and balancing of opposing tendencies, along with the avoidance of 'one-sidedness' (N.B.), is, I believe, of central importance at more than one level of personality development, and initial design. Each of the poles of the emotional compass discussed earlier has in the (notional) full personality an equal strength – that is, if 'male' and 'female' are somehow brought together into one, we have, instead of two situations dominated respectively by love–hate and fight–flight, one stable configuration where the four forces hold each other in balance. While the terms I use here for the four poles are other than those employed by C. G. Jung, one has here something in form very like his *mandala*. A mandala is any harmonious set or pattern of four objects – which Jung found in a variety of paintings, dreams and stories. He held also that towards the end of successful psychoanalysis such images would tend to occur in the dreams of the patient. Essentially my views here are not at variance with those of Jung, despite the considerable differences of detail. However, I consider the mandala to be not an archetypal, but in my terms an archestructural phenomenon.

The Self and Ego are alike notable for their lack of humour. So one may search the Bible, the *Koran* or the *Upanishads* on the one hand, or the stories of heroes, the history of nations and scientific textbooks on the other, and discover virtually no intentional humour. The Self and Ego do not, and cannot, laugh themselves – though they occasionally mock each other. For the releasing and relaxing experience of true humour one requires, I suggest, both to comprehend, and to be outside, both Ego and Self.

The subject of morality is one I shall be developing a little further in this volume. I have elsewhere suggested, however, that morality can be based only on a true freedom of choice – and that such freedom does not exist within the confines either of Self or of its partner, the Ego.

Our legends, literature and social structures abound in groups of three. Some of these are: the Holy Trinity; Freud's id, ego and super-ego; the three Wise Men; the three wishes of fairy-tales; the 'third eye' of mysticism; the thesis, antithesis, synthesis of logic; the Platonic trilogy of appetite, passion and reason; the body, mind, soul of everyday speech, and so on. So widespread is this practice of formulating theories in threes that I must consider it an archestructural manifestation. While the threes I have cited are sometimes very different in their precise content, their general and shared implication is clear. A third thing overrides, and usually arises out of, two earlier propositions. I take the latter to be Systems B and A. The outcome of their interaction is what I term System C.

Before going on to speak further of this third System, I should emphasize that while every individual perforce and automatically possesses a System B and a System A, this possession in no wise guarantees the emergence of what I refer to as System C.[35] There are a good many, I believe always undesirable, ways of coping with our 'divided selves', to use R. D. Laing's term. The most common is by compartmentalization. In a variety of ways one 'renders unto Caesar the things which are Caesar's, and unto God the things that are God's' (the most damaging notion, I think, to which Christ ever gave his support). So we, for example, love, or mix with, certain categories of people and not others; are honest towards family or friends, but dishonest in business; 'fuck' prostitutes but 'respect' our wives, sisters and mothers; accord in the view that business and pleasure don't mix; strive to 'preserve appearances' (presumably at the expense of some other aspects of reality), and so on. The catalogue is virtually endless. Again and again, despite the popular saying, we agree in practice that what is sauce for the goose is *not* sauce

---

[35] Of course, I am here speaking rather too simply and in generalizations. Most individuals achieve occasional or small-scale System C states briefly and in particular areas. What I am considering above is the achievement of a permanent personality integration on virtually all levels and in virtually all contexts.

for the gander. The price we pay for the 'convenience' of this split, in mental and physical misery, and in human lives, is of course incalculable.

In other cases, instead of tolerating a split or a house-divided situation, individuals seek to solve the dilemma by granting (virtually) full government to one or other system, or even sub-system. Some take refuge in religion (withdrawing as completely as possible from *this* world) or, say, communism; others bury themselves in business, or become fascists, priding themselves then on being realists or without sentiment. All such individuals are considered, nonetheless, relatively normal in our society. In *Total Man*, however, I suggested that the extremer forms of these *same* behaviours are *clinical neurosis* and *clinical psychosis* respectively. It is in those extreme forms that one may properly perceive the dubious value of the milder forms of the same 'solution'.

One last point here. While I have so far spoken of A- and B-dominant behaviours in absolute terms, they have also a relative aspect. I have stated that one way of living with the two systems is to make one or other as far as possible truly dominant, and I suggest that this is one significance of the term 'totalitarian'. This attempt succeeds up to a point. But still *some* concessions must be made to the rejected system, especially if this is relatively vigorous. Differing degrees of concession in the 'balance' achieved (not, of course, true System C balance – merely a working balance) must be made. Thus we have System B personalities and behaviours which are in effect closer to System A than some others (and vice versa for System A in respect of System B).

When we stand outside both systems we are able to view and label them in absolute terms as definitely B or A. But if we are looking at any behaviour from within any (other) System A, or B, situation, some System B behaviours may be *perceived* as System A behaviours, and vice versa, *because they lie further in the direction of that other System than does our current standpoint.*

This is better understood from the following analogy. If I am standing somewhere in the middle of a row of large cards which are painted blue on one side and yellow on the other, all cards in one direction appear blue, all cards in the other yellow. If I now change my position by walking in either direction and look once again, some of the cards I formerly saw as yellow (or blue) I now see as blue (or yellow). (See Figure 9.) In this 'within' situation, cards are yellow or blue depending on the standpoint from which they are viewed.

Communism, viewed from the standpoint of religion, is *not* a System B behaviour (which it nevertheless is in absolute terms), but a System A behaviour – worldly, materialistic, cognitive, and so on.

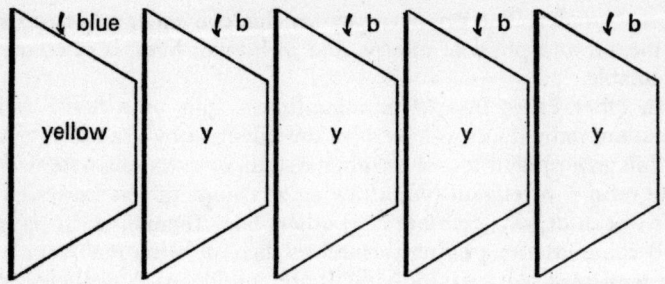

FIGURE 9
Diagrammatic representation of the differing 'values' of objects
(and situations) viewed from different standpoints.

The fascist or the capitalist, however, from his position sees communism as the System B behaviour it is – to him impractical, emotional, and so forth. Religion, Karl Marx said, was the opium of the people – but to the fascists this is a perfect description of communism.

This relative aspect of the theory is an extremely important one.

To revert, however, to System C, there exists here as with the earlier systems a unifying persona at the psychological level. I have termed this the 'Person'. The synthesizing triad is then:

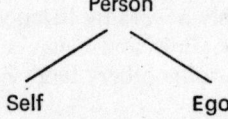

a pyramidal structure which can be extended to take in also our earlier formulations. (See Figure 10.)

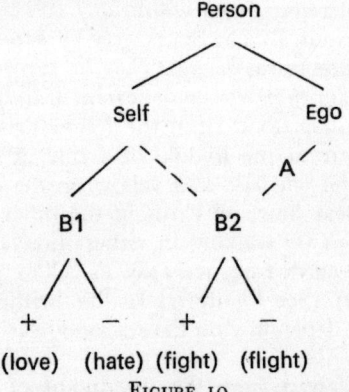

FIGURE 10

NOTE: the precise connections between B2 and the Self remain a theoretical difficulty – hence are connected by a broken line.

This, essentially, completes our review of matters necessary to an understanding of the present book. It may in conclusion be useful, however, to summarize some of the major characteristics of the three main personality systems, B, A and C – Self, Ego and Person. In the case of the last, unfortunately, we lack adequate terms for some of the concepts involved (and even the concepts themselves). One is tempted merely to keep repeating the word 'balanced'. That the requisite terms are as yet lacking in our language is one of the factors which leads me to believe that this is a system still very much in course of evolution.

Self	*Ego*	*Person*
'visceral'	'muscular'	—
'emotion'	cognition	art
female	male	—
unconscious	conscious	expanded, full or third consciousness
subjective	objective	integrated
mystical	public	—
irrational	rational	reasonable
associative	reductionist	synthesizing–creative
cyclic	linear	? triadic
yielding	dominating	balanced
'neurotic'	'psychotic'	sane

I shall be arguing in the present book that no genuinely enduring, developing or meaningful society is possible in terms of either the Self or the Ego alone. The Ego, in particular, has gone as far in taking control of the human personality as we dare let it go. Whereas under the Self there would perhaps be the possibility of some kind of existence – albeit, I regret to suggest, a vegetative and primitive one – under the leadership of the Ego there is every chance that our species will cease to exist. I believe that what we are currently witnessing in modern western society, and especially among our young people, is partly a revolt against the Ego. In this I am in the company of such writers as Charles Reich, Timothy Leary[36] and others, but, unlike them, I view much of what is taking place as nothing better than a return to the rule of the Self. This is an almost equally undesirable alternative. I shall, however, try to indicate in one of the later chapters of the present book something of what I consider to be the nature of the genuinely viable Alternative Society.

---

[36] *The Politics of Ecstasy* (MacGibbon and Kee, London, 1970).

II

PREVAILING WINDS OF THE NERVOUS SYSTEM

1
Archetype and Archestructure

In the course of his research into mythology and folklore, the psychiatrist C. G. Jung came upon what he termed 'universal images', that is, recurrent figures – the Wise Old Man, the Virgin, the All-Mother, and so on – in the expressive repertoire of different peoples in widely separated parts of the world. Believing that cultural dissemination alone could not account for the global distribution of these stock figures and situations in art, literature, religion, and perhaps notably in dreams, he therefore came to argue a genetic component in each, one common to all members of the human family, acquired in the process of our biological development as a species. These alleged universal experiences of the species form the basis of Jung's concept of the collective unconscious. He termed the experiences archetypes.

There are points of disagreement between Jung's general psychological theory and my own which need not be raised. The notion of the archetype, however, is one that I by and large adopt, but with certain additions. I myself see the Jungian archetype as closely related to, if not indeed identical with, the sign-releasers described by ethologists – although these, as described, are stimuli, while the Jungian archetype, as described, is a response. In respect mainly, though not exclusively, of infra-human organisms, ethology has demonstrated that the presence of certain constellations of events produces an immediate, stereotyped response in all members of a given species (or genus, and so on). Thus, for example, the young of game birds will at once run for cover if the silhouette of a hawk is passed over their heads; the fledgelings of many species of birds will instinctively open their mouths if the nest they occupy is vibrated; baboons will shelter from rain if the 'rain-call' of the species is played on a tape-recorder on a blazing hot day; and a young human baby will automatically smile at a piece of cardboard with two small black circles painted on it. Examples run into many thousands. I believe that the physiological/psychological reactions which occur in the perceiving organism in the presence of the appropriate releasing (or inhibiting) stimulus are the basis and model for an archetypal, mental 'experience' of the releaser or inhibitor. Of course, with regard to animals we can have no direct access to any alleged psychological or internal experience allegedly further associated with their overt, behavioural response to the releasing stimulus. We do, however, have such access to human mental states.

These are among matters to be taken up again elsewhere; but before now moving on to further consideration of my concept of the archestructure and its effects – the main purpose of this chapter – I would like to make one or two more, for the moment unsupported, remarks about the archetype, as I myself conceive it.

I find it appropriate to distinguish between 'shaping', 'living' and 'dead' archetypes. There are no abrupt breaks between these three forms of the phenomenon. Rather, one gradually becomes another. The first-mentioned deserves, and will receive, a whole chapter devoted to its nature. The last-mentioned of the three is the most readily described, and refers to a figure or situation which has entirely lost its power to arouse even a residual response in us, one which has therefore ceased to evolve – is incapable of affecting both our, and hence its own, evolution. It is no longer in the process of being set up or of adding to its evocative strength, but is now fixed. This is not at all to say, however, that such an archetype cannot undergo further change – but this now is the change of decay, loss or distortion. The idea is perhaps best conveyed by analogy. A fossil, even once formed, may be distorted by the pressures and movements of rock above and around it. So we find, for example, that the gills which appear briefly in the life of a developing human embryo are not identical in form and other points of detail with the gills which some very distant ancestor of ours once actually possessed and employed.

A living archetype – to turn to the middle category – may be conceived of as one that in some way affects, or can affect, the behaviour of living individuals, though at the same time not necessarily at all stages of an individual's development. An archetype still alive in a young child, though dead in the adult – for example, adults do not instinctively smile at a piece of cardboard with two black circles painted on it – may still be effective as an evolutionary force, for instance, in the example given, if the smiling behaviour of the young child in any way caused the parent to want to have further children. These, naturally, would tend also to possess and exhibit the behavioural response of the first child.

The shaping and living archetypes, as I shall hope to show, are in fact of crucial importance in the total evolutionary process. In passing I would here propose only that a shaping archetype may perhaps be identified, and as such relatively distinguished from the living archetype, by the degree to which the ideal, the *maximally evocative form* of the archetype, differs from the form *as actually found in nature*.[1]

When the appropriate stimulus situation (what I term the archetypal situation) occurs, we respond to it – that is, if it is still at least a partially living archetype that is involved. So the baby smiles at the black circles; so we feel protective when we see a young child; so we may make ourselves smaller (cringe) in the presence of an angry authority figure, and so on. That is on the one side. But in addition to this I suggest that, by reason of the fact that these archetypes and archetypal situations, be they living or dead or shaping, are somehow 'on record' within us, we will often also 'find' them when we are generally casting around for ideas or explanations, when creating some work of art, or indeed when ever mentally active in any way. Note what is said here. We are not *responding* to an actual instance of the archetype in our presence. We are projecting it into some vacant, or at any rate malleable, situation. On this basis archetypal forms should appear fairly regularly in our art, projections and other behaviours – and do, as Jung well noted.

---

[1] There is yet a further side to these general issues. In purely experiential as opposed to evolutionary terms – and now employing the notions shaping, alive and dead in their more everyday sense – an archetype may, as already suggested, be dead (or fossilized) in the minds of the adult members of a species, while still living in the minds of the young of the species. The notion of an actually re-experienced recapitulation by the young of at least parts of the biological history of the species is then involved here – not, of course, one to which all psychologists subscribe. It would then be possible to argue that physical adults, who still actually experience as living, archetypes only legitimately living in the young, are very literally arrested or fixated at some earlier stage, at least in that particular regard. This would give some literal basis to otherwise largely notional concepts of arrest and retardation in psychological theory. It is further possible that in some forms of mental illness dead archetypes are re-activated, so that the patient experiences them with the immediacy and reality that once did our adult ancestors (and/or as children do in the course of growing up). Of course, the physical mechanics of the processes just envisaged pose some fairly formidable problems at the physiological level.

It goes, or I think should go, without saying that a given archetypal response, even with all other things being equal, will not occur in the same strength in every member of a particular species. I suggest that, statistically, we have some kind of normal distribution of response strength, in any reasonably large sample – just as we find for height, weight, running speed and many other hereditarily determined functions and attributes.

In the present chapter, however, it is not primarily the archetype but the archestructure that I wish to discuss. The archestructure is my own formulation, not entirely unrelated to the archetype – hence my choice of a deliberately similar term. I define an archestructure as a somehow perceived or experienced feature of the structure or functioning of the nervous system, which is then 'discovered' in (i.e. is projected or injected into) our physical or psychological surroundings : that is, may be found in our products, inventions, descriptions and social behaviours. The term 'nervous system', incidentally, must be taken when used by myself to include also the *evolutionary history* of our nervous system.

This definition of the archestructure is not wholly inappropriate also to the archetype – and in a sense that concept may be thought of as a special case of the archestructure. But there are as well many differences between the two. I believe that the archetype is evolved or acquired as a (biological) response to a quite specific set of circumstances in an organism's evolutionary history – the acquisition of which confers survival value. Thus the archetype is essentially a fairly small-scale phenomenon and associated, in terms at least of its acquisition, with a relatively short span of time.

The archestructure is in general a much larger-scale phenomenon – so large, in fact, as readily to escape notice. One fails, in other words, to see the wood for the trees. This considerable scale aptly reflects the fact that not merely a few billion cells of the cortex are concerned in the generation of the phenomenon (as I suggest is the case with the archetype) but entire sections of the total nervous system. While I believe science as a whole is archestructural, as are religion and the party political system, it is not only in these larger arenas that archestructural influence is seen at work. There are archestructures of considerably lesser range, though nonetheless impressive for that. The archestructural impulse, then, does not disdain to work through the most trivial and matter-of-fact situations of everyday life.

I have no idea what first caused me to register that in the large majority of houses and hotels I visited the hot tap was on the *left* of the basin, not on the right as one might perhaps expect – if indeed the arrangement were not to be wholly random. A call to the Building Centre in London after some delay elicited the information that to make the left tap the hot appears indeed to be standard practice – but nobody could tell me why.

My own suggestion is that what we see reflected here is at once the 'left-handedness' of the cerebellar system, the Self and the unconscious generally – and the emotional warmth of those systems – while the opposite tap appropriately and conversely bears the association of the unemotional (cold here = calculating) cerebral cortex, System A and the Ego.

On a grander scale basically this same phenomenon is seen in the well-known (though so far unexplained) fact that in many conurbations the poor live in the east of the city, the wealthy in the west. This is the case in both London and Berlin, for example. My own construction is that in general the poor are B-dominants, the wealthy A-dominants. Arriving at or being born into the city as more or less a mixed bag (at any rate as a random series), the mixture then almost from the first nevertheless 'settles out' into its component parts (as, in the chemical world, oil will from water, or cream from milk). The settling-out – because not controlled or directed by other unconscious or reality factors – occurs along the 'lines of force' of the general polarity of the human nervous system. The East, where the sun rises, is associated with the beginning or birth of life (the paleoanthropic B-dominants are the older in evolutionary terms – hence also such phrases as 'the mystic East'). The movement of the sun towards the West on the other hand somehow, I suggest, evokes or parallels the questing nature of the A-dominant. The neanthropic is on the way to becoming. In this general connection – though this is not a matter on which I wish to expand in the present book – I have noted that the migrations of A-dominants are normally to the West, those of B-dominants towards the East. The 'Go West, young man' of evolving America was then a quite specific, though of course unconscious, reflection of this age-old situation.[2]

I should emphasize at this juncture that the influence of the archestructural impulse, while it may be thought of as a kind of giant Freudian slip, is essentially not a neurotic or disturbed one in the clinical sense. Nor am I suggesting that it cannot be overridden. It is when there is a relatively free or unstructured situation, on which no reality or other determinants happen to be operating, that the gentle, though very insistent, push of the archestructure is manifested. Just as the subject of a Rorschach test finds his own personality patterns in the ink-blots – because there is nothing else for him to find – so when a general situation is otherwise 'free-floating' it will, I propose, tend to constellate into archestructural forms and patterns. The archestructure (like the archetype) is *par excellence* among what I term *the prevailing winds of the nervous system*. We can readily erect barriers against them, we can fairly readily tack into – that is, sail against – them. But they blow and never cease to

---

[2] The ancestors of the Red Indians, of course, had gone East from China into Canada and thence the Americas where, alas, they again met their ancient – i.e. evolutionary – adversaries coming the other way on our, unfortunately, round planet. 'Cowboys and Indians' are a further specific rendering or encapsulation of the general Cro-Magnon/Neanderthal struggle.

blow. And whenever we let fall the tiller, or fail to repair the barriers, these prevailing winds once again show their untiring influence.

The foregoing has possibly suggested an all-or-nothing situation – that is, that the influence of archestructural forms is either solely dominant or else not present at all. This is not so. To revert to the analogy just used, if a windbreak has several holes in it, the wind will blow through it to that extent. Or if it has just one tiny hole, the wind will nonetheless still blow through that as best it can. A situation which is very largely reality-determined (i.e. by external or so-called objective reality) may still possess 'free' aspects which the archestructural tendency can utilize. Something of such a situation we saw earlier with our hot and cold water taps. Buildings are *otherwise* planned to the last detail. In short, human behaviours and products may be archestructurally influenced or qualified on a scale from zero to one hundred per cent.

While the archestructures of both the Self and the Ego (particularly in the case of the latter) can and do exist in virtual independence of each other, the composite Self-Ego archestructure is very wide-spread. The consideration of such composite, dual archestructures occupies the remainder of this chapter.

Much of the received basis of the theory and practice of classical Greek drama is summarized in the twin masks of comedy and tragedy. Versions of these indeed still adorn the facias of most modern theatres, sometimes too the stage surround, and often the printed programme. I believe these figures to be shorthand archestructural descriptions of the essentially dual nature of mankind, of the Self and Ego respectively. And if additionally we now consider the stylized renderings of a representative paleoanthropic and a representative neanthropic skull in Fig. 11, shown alongside the actual skulls, do we not have something remarkably like the two theatrical stereotypes?

Some rather scattered observations here, whose general relevance will be nonetheless apparent. When referring to a sad person we often say he has a long face. The neanthropic face, as it happens, is long, the paleoanthropic short. Moreover, fat people (round people) are popularly supposed to be jolly and fond of good living. Long thin people are habitually associated with sadness, seriousness or too much thinking. We think perhaps of Falstaff[3] and Sir Toby Belch[4]

---

[3] Shakespeare, *King Henry the Fourth*, Parts I and II.
[4] Shakespeare, *Twelfth Night*.

Archetype and Archestructure

in the former respect; of Malvolio[5] and Cassius[6] in the second.

A tragic figure is one who falls or is brought from some high estate to a low estate. This movement may be literal in the sense that an office (say, that of king or commander) or great wealth is forfeited; or metaphorical – when great happiness turns to great sadness or suffering, for example. It is of the former that we speak first.

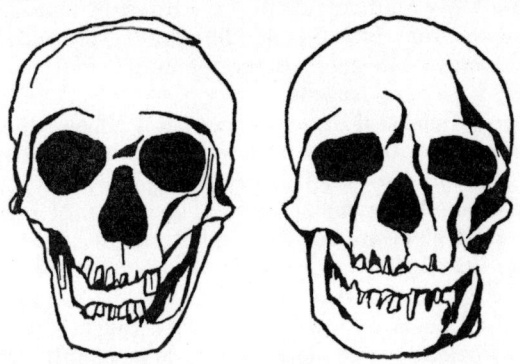

(a) Skulls of Neanderthal and Cro-Magnon. The teeth and mouth of Neanderthal are larger and more protuberant in a generally more muzzle-like jaw. The mouth of Cro-Magnon by contrast is in all respects smaller and tighter. Neanderthal has larger, rounder eye-sockets and a larger, rounder nasal cavity. The downward slanting eye-sockets of Cro-Magnon are characteristic and one of the main respects in which this skull differs from that of modern man.

(b) Stylized representation of the main features of the two skulls. Note their 'coincidental' similarity to the traditional masks of comedy and tragedy.

FIGURE 11

---

[5] Shakespeare, *Twelfth Night*.

[6] Shakespeare, *Julius Caesar*. 'Yon Cassius has a lean and hungry look; he thinks too much: such men are dangerous.' And: 'Let me have men about me that are fat ... and sleep o' nights.'

Greek classical tradition held firmly that only the great or noble individual could be the subject of tragedy – preferably a king or ruler, but also certain members of the upper crust. Only the *great* fall could be tragic: hence the common man could never be the subject of such drama. He was properly the subject of comedy.[7] The classical Greek view was also followed in the later French and German theatre. Only gradually from the eighteenth century on is the notion accepted that first the middle classes (and ultimately the working classes) are fit subjects for the tragic form – leading to the so-called 'bourgeois' tragedy.

The theory of classical drama reflects, as I believe, the historical situation of the conquering of a despised people (the Neanderthaloids) by 'superior' Cro-Magnon between 35,000 and 25,000 B.P. (i.e. Before Present). In the subsequent amalgam of these peoples Cro-Magnonoid elements are concentrated in the aristocracy and the military (particularly the 'officer class'); Neanderthaloid elements in the agrarian and urban masses. Social theory and social structures of all kinds (including the theory and structure of Greek drama) unconsciously – or more or less unconsciously – reflect this situation.[8]

A word on the nature of the comic or homely figure. As I have already suggested, it is incorrect to assume this individual to be only and always light-hearted. On the contrary, the sad or downcast aspect of these figures is frequently shown. So such phrases today as 'the broken-hearted clown', 'laughing on the outside, crying on the inside'; so the great aria 'On with the Motley' in Leoncavallo's *Pagliacci*, and also too the reverse side of the later Falstaff.

Indeed I believe the essential nature of the paleoanthropic or B-dominant to be manic-depressive – a continuous oscillation between joy and sorrow. Of course, the phrase manic-depressive refers properly to a serious clinical psychological condition. It is the normal and milder form of this condition, although much more clearly visible in mental illness, which I have in mind here.[9]

---

[7] The term 'comedy', it should be emphasized, does not exclude a certain seriousness. I note with interest here that some philological authorities consider 'comedy' to be from the same root as 'home'. 'Homely' would be an excellent term to apply to this type of drama. Compare with the 'noble' content of tragedy.

[8] I am impressed, too, by such parallels as the fact that 'status' and 'stature' are from the same root, and that stature of course means height. Cro-Magnon was on average much taller than Neanderthal.

[9] The Ego prides itself on its stoicism (constancy, steadfastness, etc.). It is contemptuous of the oscillatory nature of the Self – which it stigmatizes as *vascillatory*. 'How can one take such people seriously?' the Ego asks. And goes on to say: 'One minute they are up, the next they are down. It

A further version of the archestructure expressed in the classical Greek divisions of the tragic and comic muse is found in the comic double-act of today's entertainment world, a widespread and much loved phenomenon. The standard version of this act consists of two men, one rather thinner and taller (the so-called 'straight' man, N.B.), and one rather shorter and fatter (the stooge). The former is supposed to be smart, the latter stupid. In fact, of course, the victory by no means always goes to the former – and it is in the interplay of the differing potentials, in the contrast of styles of the two men that the interest and amusement are centred. Well-known examples of such comedy duos are Abbot and Costello, Olsen and Johnson, Laurel and Hardy, Dean Martin and Jerry Lewis, and Morecambe and Wise.[10]

We can see that these duos are, fairly literally, man and wife – certainly 'male' and 'female'. Men are in general taller and slimmer than women, women in general shorter and rounder than men. This pair is once again also Cro-Magnon and Neanderthal, Ego and Self, respectively.

The situation we have been examining in these last paragraphs is further duplicated and extended in the symbolism and phraseology of 'cat and dog'. A stressful marriage is frequently described as a 'cat and dog life'. The term 'dog' is regularly applied to males (as in dog days, and old dog, a sea-dog, a salty dog, a gay dog, and so on), while a woman is often a cat (hell-cat, cat-house=brothel), and her alleged habit of putting down other women is termed 'cattiness'. And while a fight among women is often called a cat-fight, an aerial combat, for instance, among men is a dog-fight.[11]

I do not believe it is claiming too much to suggest that in the course of a good many thousand years mankind has acted (largely, though perhaps not entirely, unconsciously) as an instrument of natural selection on the two animals in question – strengthening the 'male' qualities of the dog and the 'female' qualities of the cat. Those individual animals who exhibited these appropriate qualities

---

is much better if I take charge of affairs generally and their affairs in particular for them: they are obviously not fit for self-government.' We glimpse here some of the roots of paternalism, masculine society, despotism and the Fascist State.

[10] It will be noted that the standard characteristics are not always rigidly shared out along the lines I have proposed. Thus Oliver Hardy, though taller, was fatter than Stan Laurel. Eric Morecambe (the stooge) is both taller and thinner than Ernie Wise, the 'straight' man. This is not too important a point – the basic position is still recognizably the same.

[11] The term for a female dog, 'bitch', is certainly also much used of women.

stood, I suggest, a slightly better chance of survival than their fellows, with the result that in time these characteristics became strengthened in the whole species, along Darwinian lines. And so the lonely bachelor keeps a dog, the spinster a cat – that is, a friend of their own sex. (The witch, too, of course, keeps a cat.)

We may mention in passing here a much older archestructure of similar composition – the lion and the unicorn. Here the lion is the male symbol. This might appear to contradict all that I have just said about the cat. But not so. We are speaking of the male lion, and the least feline of all the cat family (his mane, perhaps, is a beard, and he is, incidentally, the only member of all these genera who possesses such), renowned for his physical strength. His names also are Lord of the Jungle and King of Beasts.[12] The unicorn here is the female element – gracile, beautiful, and so on. The single horn is not a phallus (though see my footnote) but, I suggest, a reference to the ancient 'one-ness' of personality, destroyed, originally, by the evolution of the cerebrum, and further in historical time by the crossing of the paleoanthropic and neanthropic varieties of man.

With that we turn to a very different area of human activity, that of music, where again however we find much evidence of paired archestructural forms. I propose, taking piano music as my particular example, that the treble (or melodic line) is the Ego, the bass (or rhythm line) the Self. There are many pointers to this conclusion. The bass is played by the left hand, for example. It is true that the more complex melodic work is *logically* reserved for the more dexterous right hand – but that constellation of circumstances is itself no accident. The repetitive rhythmic bass originates in the percussive drum-beat, which in turn I am not alone in suggesting derives in some sense from the pulse beat of the human heart.[13] 'Beat' and 'rock' music has a much stronger emphasis on this rhythmic aspect of the whole than has the sophisticated, melodic, music of Europe. Beat music in the broadest sense of that term is the preferred music of many B-dominant peoples – Negroes, and so forth.

There are here many further aspects both fascinating and germane. An alternative spelling of bass is base. Skeat[14] derives the word from the *feminine* form of French *bas,* meaning low. Both

---

[12] This, being a much older artifact – a legend, actually – is nearer to a dream than it is to waking reality, and the figures are almost entirely symbolic. As (dream-) symbols there is in any case nothing to prevent both of these figures representing both parts of the equation at one and the same time.

[13] Cf. Desmond Morris, *The Naked Ape* (Jonathan Cape, London, 1967), Chapter 3.

[14] *An Etymological Dictionary of the English Language* (Elsevier, Amsterdam, 1967).

Skeat and Klein[15] connect these words in origin to Latin *bassus* meaning short, fat, stumpy (shades of Neanderthal!), base, of course, in any event meaning also low, underhand, sly (a clear reference to the nature of women and cats, as every dog would agree). And which is the 'underhand'? It is I think not necessary to spell out all the double meanings and interconnections of these various terms.

There further exist in our music the major and minor keys. Without entering into the full technicalities, the minor key is described as a 'diminished' form of the major. These two keys are, I suggest, whatever else, yet further archestructures of the Ego and Self respectively. Many of the plaintive songs of the Jews and other oppressed peoples are written in the minor key.[16] And here we touch on other behaviours of the Self related to these general issues. The Self likes to sing its suffering – compare additionally the Blues tradition of the American Negro. The Ego, ideally, bears its sufferings in silence and without complaint. Again ideally, it does not break this silence even under torture. It does not request mercy. Nor does it *request* freedom – it simply takes it, or dies in the attempt. 'Let my people go' is the *request* of the *Self*. (And one must report, sadly, that those who merely ask for their freedom do not usually get it.)

There is, indeed, a sense in which the Self does not want to be free and in which it enjoys its suffering. This is a statement one must make with caution – for, though it is, I believe, in a certain sense true, it is the kind of statement which is apt to be only too useful as fascist propaganda. Is one saying, for example, that the Jews enjoyed the concentration camps of the last war? Or that the Negroes wanted to be shipped as slaves to America and elsewhere, there often to die of starvation and mistreatment? That obviously goes much too far. Yet there is a sort of truth here. There is a joking question which asks: 'What does a Jew fear more than persecution?' to which the answer is: 'Being overlooked.' And something of this is true, I think, also of women. Some women, at any rate, seem willing to accept a good deal of physical violence and other ill-treatment from the men they love. The wanting of punishment or suffering has in fact obtained clear recognition, not only in clinical psychology, but in common parlance, with the term 'masochism'. A masochist is a person who, in some sense of the terms, enjoys suffering and invites pain.

---

[15] *A Comprehensive Etymological Dictionary of the English Language* (O.U.P., London, 1966).

[16] In a television programme in June 1972, Hans Keller, the music critic, commented that the three great Jewish composers (Mendelssohn, Mahler, Schoenberg) use the minor form far more often than non-Jewish composers.

In the equally popular term of 'sado-masochism' (sado- deriving from sadism) we have once again the full Ego–Self equation. It is the Ego which takes pleasure in inflicting pain, the Self that takes 'pleasure' in receiving it. Full sado-masochism is, strictly, once again a clinical condition or conditions. I believe personally, however, that mild forms of this dual behaviour are both widespread and normal in our culture; and I would argue the existence of this condition to derive, ultimately, from the basic sexual behaviours and responses of the mammals as a whole class. Thus we note, for example, that the male cat, guinea-pig, and so on bites the back of the neck of the female during sexual intercourse – and is not 'permitted' to have intercourse by the female unless he does this : that is to say, this action is a releaser in the ethological sense, necessary for the triggering of certain automatic responses in the female.

To revert, however, to the minor key and the 'celebration of suffering' in B-dominants. Not merely in, say, moments of reflection, but at funerals and in religious ceremonial generally, the Self vociferously and often hysterically indulges its suffering. I am not trying to say that there is no genuine sadness at the loss of a loved one at a Jewish or Negro funeral; but there is, too, a relishing and a treasuring of the suffering. This is enshrined elsewhere also, for example in the paradoxical phrase 'a good cry' – and women sometimes actually speak of 'enjoying a good cry'.[17]

With these remarks I conclude the detailed consideration of the

---

[17] I would like in passing to emphasize what I consider to be the difference between hysteria and frenzy. I am aware, of course, that many people use these terms more or less interchangeably. Hysteria is, for myself, properly applied only to the excesses of the Self – cf. also the hysterical paralysis of Freudian neurosis. Frenzy I use to describe the excesses of the Ego – for example, in going berserk, running amok and so on. Frenzy frequently involves killing, and is by no means as unco-ordinated in terms of its actions as might appear at first sight. The Nazis, then, were not hysterical, but frenzied. An over-enthusiastic supporter of Women's Liberation is not frenzied, but hysterical. I suggest that hysteria is a loss or disintegration of (coherent) response; while frenzy is uncontrolled, exaggerated *action*.

Technically, I take hysteria to be 'sexual' energy (which arises in the Self) present in excess, and usually in addition incorrectly or inappropriately directed. As Freudian theory maintains, I maintain also that the blocking or failure to use this emotion (in Freudian terms, libido) for its rightful purposes causes it to become explosive, not wholly unlike a head of steam, and/or to be diverted into whatever substitute channels are still allowed it. Frenzy I take to be 'fighting' energy (which originates in the Ego), also present in excess, similarly blocked, hence similarly explosive, and likewise capable of misdirection into substitute channels or towards inappropriate goals.

composite Self–Ego archestructure. I have not exhausted the possibilities nor, deliberately, recapitulated the many examples already discussed in *Total Man* – the ventriloquist and his dummy, the puppeteer and puppet, Svengalis and their Trilbys. I should perhaps, however, point out once again that the party political system is one of the clearest (and largest) examples of the antagonistic archestructural process in our society, by its action influencing and shaping vast areas of our public behaviour and public spending and (wholly regrettably, I think) absorbing untold quantities of human energy for which better use could be found. The interaction of Self and Ego archestructures further plays a major role – this time non-antagonistically and, as far as I am concerned, to wholly worthwhile ends – in all forms of art and artistic creation. To this central issue a later chapter is devoted.

The next chapter, however, is given over to a consideration of the role specifically of the Ego-archestructure and of certain Ego-archetypes in modern Western civilization.

2
Ego-Archetypes and Ego-Archestructures in Western Civilization

(i) Ego as Ego

In the previous chapter we were concerned with the archetype and archestructure in general, and the composite archestructure in particular. In this chapter I shall be considering the virtually 'free-standing' Ego-archetype and Ego-archestructure. The word 'virtually' is important.

This is perhaps very much a point at which to remind the reader that each one of us – that is, each and every member of the species *homo sapiens* – is *always*, and at all times, a combined Self and Ego. (This holds true even for the extreme psychopath who is, as I consider, abnormally underendowed in respect of Self and System B qualities.) Though I continually speak of these as two contrasted and antagonistic elements, which they certainly often are, both are nevertheless housed within, and in combination substantially form, the organism known as man. Each of us, then, is Faust. In each, two spirits live more or less unwillingly in more or less unity. Even though at times, or in general, one aspect of our personality may predominate, the influence of the other is never far away, never wholly undetectable. I shall be pinpointing the precise whereabouts and form of the subordinate element in several of the instances we consider.

One other background comment concerning general theory may be conveniently made here. It is possible – and, for the time being at any rate, advisable – to think not only in terms of specific Ego-archetypes and Ego-archestructures, but also more loosely of general ego-behaviour – of simple assertiveness, say, of what are termed egoism, egotism, and so on. I am inclined very much to argue that ego-behaviours will *in general* and even in principle be based upon, or derive from, some specific and specifiable Ego-archetype, archestructure or archetypal situation, even though this underpinning may not be all that clear at first sight. That the large part of ego-behaviour *is* so based I do not doubt. Yet I feel a slight reticence at attributing all ego-activity unreservedly to the influence of archetypes or archestructures. Hence for the time being I would wish to retain the interchangeable terms of ego-behaviour and ego-activity as simple general referents.

Reverting to my earlier point – that Ego-products generally seem to show residual traces of the Self, and that though seriously diminished the influence of the Self is never entirely absent from them[1] – I believed originally that this held for the Ego-archestructure, but not for the Ego-archetype, and that in realizing this I had found both a useful and reliable means of distinguishing between those two phenomena. Thus, when I originally considered the game of chess (to jump for a moment a little ahead of ourselves) I took this to be an Ego-archestructure, showing traces of the Self, heavily and specifically overlaid with a combat arche*type* – a strictly Ego to Ego affair. At the *first* level, I considered that game to be an archestructural statement of the evolution of B2 into A.[2] The pieces are mainly warriors; and the basic situation is one of battle (B2). Yet the conceptual possibilities of the game astronomically exceed those of any real battle – and indeed the game is one of the highest tests of human ingenuity and concept-span (System A). And yet, in this clear province of System A and the masculine Ego, there are two

---

[1] Certainly, that is, as long as the behaviour in question is still in the range of what we term normal. At the point where the Self is wholly excluded by the Ego – the point at which the Self should (on the Ego's expectation) vanish completely – a sudden, dramatic change occurs. I prefer for reasons which will be obvious to term that moment 'break-up' rather than 'break-down'. For at that point the (frustrated?) Self suddenly abandons all attempts to live with the Ego – as it were breaks off diplomatic relations – and appears (or continuously and imminently threatens to appear) to the Ego in all its terrible separateness – that is, as the hallucinations and terrors of psychosis and madness. Far from having been obliterated or destroyed (by the Ego's action) the Self rises up more than ever as an independent force in its own right.

[2] As I argued in more detail in *Total Man,* I hold (logical) thought to be highly evolved muscle, just as I take emotion to be highly evolved viscera. Instead of literally having to perform an action with movements we became able to perform it symbolically in our heads. Together, of course, with much actual experience of the world of physical objects, we are able to visualize, too, the (probable) consequences of the actually unperformed action.

B-figures – the bishop and the queen. These, I suggest, symbolically represent the Self.[3]

So I reasoned the game of chess. But, as we shall see, traces of the Self are found also in what seem to be *purely* archetypal situations, where there is no apparent archestructural underpinning. Consequently I am for the moment once again persuaded that no absolutely final distinction between archestructure and archetype is possible.[4]

As a reasonably separate or uncomplicated example of the Ego-archetype we may take first what I term the target archetype. This consists, essentially, of a desire to hit, and a satisfaction in hitting, a set mark. It is extremely widespread (and I take its popularity still today as an indication of its once great importance in our phylogenetic past), it being found, in isolation, variously in sharp-shooting (rifle, bow, etc.), golf, darts, bowls, bowling, the coconut shy and other fairground amusements; also as an important part of games like football, rugby, hockey, and so on, and of such guessing games as 'I Spy'; and also metaphorically in factory production or plans for social progress and reform – where one regularly speaks for instance of 'production targets' and very frequently of one's 'aims'. I have suggested elsewhere, and emphasize again here, that the appropriate participation in archetypal situations (as with the gratification of instinctual urges of all kinds, to which they are related) gives pleasure, often a considerable degree of pleasure, to the participant. Thus, scoring a bull's-eye in darts for instance is both extremely gratifying to oneself and an achievement acclaimed by others.

I take the target archetype to have arisen during man's (principally, however, the neanthropic's) days as a hunter. At various times of scarcity of game and other food, the ability to hit and kill an

---

[3] The queen and her role in chess throw up a very real theoretical difficulty. All kinds of considerations demand that the roles of the king and the queen in chess should be exactly opposite to what they in fact are. The *king* should be the most mobile and strongest attacking piece, defending his queen. But it is the queen who defends the very helpless king. I have no explanation for this unexpected turn of events, other than to plead, rather weakly this time, that symbolic, not objective, reality is involved. I am anxious always to avoid the charge of special pleading – and it would seem quite unreasonable to read meaning into what others would call chance situations, as I habitually do, but shout 'chance' when the meaning happens to go against me.

[4] Or, actually, necessary. We are not here in the domain of the physical sciences, where watertight distinctions and sharp points of cut-off do in some local sense exist, and can, within those terms of reference, be insisted upon.

animal (be it with a stone, a wooden spear, or whatever) must frequently have meant the difference between living and dying. And not, of course, just hitting, but hitting in a vital, disabling spot – the heart, the head, the eye? For, in general, you had one chance and one chance only. After the first attempt, successful or otherwise, the herd or flock was gone. Those individuals who were best at hitting the target, and hitting it first time, survived best. And those who derived pleasure from so doing were in that doubly advantaged – for they, I suggest, gladly spent much of their spare time improving their native throwing skills and, no doubt, also in refining the nature of their throwing weapons.

A second instance in our series is what I term the competitive archetype. This is likewise extremely widespread. All games involving two individuals or two teams are instances of it – and we note that the two-sided game is universally preferred to the three-sided, five-sided or whatever, game which exist as conceptual possibilities and are, I believe, occasionally found, for example in China.[5] The precise origins of this powerful archetype are less easy to understand than those of the target archetype – though I shall try to argue these through in detail in a subsequent volume. It seems that at some point doing better than an opponent in a face-to-face situation conferred survival. It is, I think, too easy to visualize the *general* combat situation – of Tribe 1 fighting Tribe 2, with the survivors living on. The origins, I feel, need to be far more personalized. I myself believe that this archetype evolved in the days when the neanthropics lived in isolated family units in a terrain that would support just so many individuals and no more. I suggest that the ritualized duel, of both recent European and American gunfighter tradition, where one competitor actually dies, shows the likely real nature (and original purpose) of the two-sided competition.

Let us consider finally in this series of instances the hero archetype. The hero is the individual who is best, bravest, cleverest. He appears in many forms – the man who survives against overwhelming odds, who slays the monster, who leads his people from captivity, who finds them new territories, or what have you. There is little need to describe this figure in detail, so familiar is he to all of us. The archetypal situation here also links clearly with the target and competitive archetypes. The hero will tend to be more highly endowed than average with most Ego qualities – that is, will be, and need to be, a good marksman, and so on.

---

[5] One distinguishes here between the genuinely three-sided game (say, having three goals) and the competition between three, seven, ten or any number of individuals, to see which *one* is best.

Where in the cases cited, however, is the Self element? My position in this respect is what common parlance terms 'Freudian' – and for some, perhaps, will prove too Freudian.

Let us take first the target archetype. That many weapons (spear, bullet, sword) resemble the human penis may, possibly, be held to be an example of parallel, though otherwise fortuitous, development. The objects which pierce most effectively are inevitably those which are fairly thin and fairly pointed. And in the case of *throwing* objects, both aerodynamic factors and the wish to strike accurately at very precise marks further encourage the trend towards streamlining, towards thin, sharp objects.

Yet we are too partisan if we imagine (a) that *hunting* is the only or even the best method of capturing game and (b) that the long, pointed object is the only, or even clearly the most effective, throwing device.

Other tactics of collecting food 'on the hoof' – which, as I believe, were more common among true *paleo*anthropics – include digging holes and lightly covering them so that animals fall into them (and later the use of nets for this purpose, notably in fishing) or driving animals over cliffs; also enticing or driving animals into artificially constructed dead ends, or into small valleys and ravines which have no exit, where they may be clubbed to death; and dropping or pouncing on individual animals from ambush (as the chimpanzee and other primates still virtually do) and biting through the jugular vein, or snapping the neck, and so on. It is clear, of course, that some of these methods and those already discussed are better or best suited to particular environments; and I am sure that terrain is additionally one of the several factors which tends to encourage, or discourage, certain hunting styles.

As regards throwing we might think of the bolas of the South American Indians – two or three rocks joined to each other by lengths of creeper, skin or rope which are whirled round the head and flung at the legs of game. They entangle with the limbs and the animal falls helpless. The lasso is a variant of the same general principal. The boomerang of the Australian aborigine, while it has a relatively sharp cutting or stunning edge, is not straight or pointed; nor is its flight path. The peoples who have evolved these weapons are, we note, effectively B-dominants.

The evolution and in particular the choice of the long, pointed object as favourite weapon is, then, not quite as inevitable as it at first sight seems. Further aspects of these weapons such as their hardness are also suspect, as we shall see later, for hardness is a feature of many other Ego-archetypes and archestructures. I suggest, therefore, that *one* of the factors, regardless of additional reality factors, favouring the choice of the pointed type of weapon is its psychological association with, and physical resemblance to, the male penis.

If finally on this particular issue we consider the various hunting styles described very broadly, it is not wholly impossible, and I believe not wholly unreasonable, to consider, say, luring and netting on the one hand; and chasing and spearing on the other, as being closer respectively to seduction and rape – that is, respectively to the way of the Self and the way of the Ego.

What might represent the Self, say, in the game of football? Apart from beating the other team, each side is involved in defending a home territory. May the goal (and the goal-keeper) represent home, wife – vagina? A far-fetched enough notion, no doubt. Yet what of the almost orgasmic roar of pleasure which comes from the spectators – almost entirely males, of course – when a goal is scored? Rugby football has no goal-keeper – and the kick is not so much into as over. Does this mean a less sexual – or more homosexual – encounter?

In wrestling, boxing and duelling I suggest, less tentatively now, that the Self is represented by the seconds. In the first two sports their comforting, supporting role is direct and obvious. The word second, of course, means the other of *two*; it is from Latin *sequor* meaning to follow – a link perhaps with the Follower?

The hero archetype, too, can frequently reveal traces of what appears again to be the Self. First, there is the not infrequent tendency to stigmatize the opponent – to describe him not as a noble equal, but as a representative of the forces of darkness and evil. In the knightly passage of arms, in honourable war, and public competition (say, the heavyweight boxing championship) there is, it is true, a tradition of *not* demeaning or 'blackening' (N.B.) your opponent. But *still* how very ready we are to downgrade him at his slightest actual or imagined deviance from the strict letter of combat or procedure (especially when we lose!). So perhaps the frequent image of the fallen knight, the 'bent' copper, the got-at jockey, the low punch, the unworthy opponent. These *precise* terms are important in their covert indication of the Self. So 'fallen' occurs also in fallen angel and fallen woman; 'bent' is the opposite of straight (one of the most reliable Ego-indicators); and 'low', of course, is the opposite of high.

I have brought out these points at this juncture mainly to show why I developed doubts concerning my notion, since abandoned, that *archetypes* exist encapsulated from the remainder of the personality, while the *archestructure*, whatever its particular emphasis, must, however residually, express the totality of the dual personality.[6]

---

[6] The matter of what constitutes the residual, or, rather, *nascent* Ego in the Self-archestructure is not a matter I especially wish to raise – but very

In turning now to look at the automobile, and later at the structures of commerce and capitalism, we encounter the compounded or complex Ego-archetype : that is, the phenomenon which presents us simultaneously with a great many separate Ego-archetypes combined into one situation. This tendency was already to some extent exemplified by the game of football, which incorporates at least the target, the competitive and the hero archetype in one whole. A heavily compounded Ego-archetype in some ways begins very much to look like an Ego-archestructure – so much so that it gives yet further grounds for doubting the advisability of attempting to distinguish absolutely between the archetype and the archestructure. There are, nevertheless, important additional features of the Ego-archestructure that are not present in the compounded Ego-archetype, as we shall see. This almost mechanical joining together of archetypes to form a functioning whole, incidentally, is a process much harder to show in respect of the Self and its archetypes. In this instance too the potentials of the two major *personae* seem, as ever, reversed.

So commonplace is the notion of the car as a symbol of power, wealth and status that it seems unnecessary to do more than mention those facts. Psychologically it is obviously very much more than a means of transport. As all seem agreed, it is in many ways an extension and exaggeration of the male ego. It is, for example, strong, fast and glamorous. I propose not to dwell on these apparently self-evident aspects, but to examine some rather less obvious ones.

A great point is made in advertising of a car's powers of acceleration – usually expressed as the precise number of seconds and fractions of a second it takes the vehicle to move from 0 to 10 or 50 or 60 m.p.h. In this connection we note that the reaction time of the male (to a stimulus) is faster than that of the female at all ages.[7] In

---

often this takes the form of some reference to light. So, for example, in the stories of vampires and others of the undead – who are otherwise in their own domain omnipotent and immortal – the item they fear and avoid, that which destroys them, is daylight. Daylight, translated, is, of course, the Ego and waking consciousness.

[7] See R. S. Woodworth and H. Schlosberg, *Experimental Psychology* (Methuen, London, 1955), Chapter 2. The authors call this 'a curious fact' – that is, one for which no ready explanation exists. I find the explanation obvious. In communities where the male hunts and goes to war, and the female does not, a survival premium is placed for males on speed of reaction. Those who react more slowly get killed more often. In due course this elimination process raises the average level of male reaction time, along Darwinian lines. The greater adventurousness of young boys over young girls probably also helps along the process.

line with my general theoretical position on all archetypes and instincts, I believe that the male derives great pleasure from the use of his reaction speed. Again the 'gun-fight' comes immediately and particularly to mind. There also exist many less drastic versions of that game (!) where the exercise of reaction time is the sole purpose – one, for example, in which one individual strikes the back of an opponent's hand very painfully with his knuckles, unless the opponent can get the hand out of the way in time, at which point the roles are switched; and many other harmless diversions are widely practised where reaction speed is a major element – in the card game of 'snap' for example.

The motor-car, in particular the sports car, by extension confers a super-fast reaction-time on its owner. The pleasure which the male takes in this manifestation is one of many reasons, I suggest, why women in general will never enjoy driving a car as much as does a man.

We turn to a very different aspect. It will be recalled that the archetype is normally described by Jung as a response, while the sign-releaser of the ethologist is, habitually, a stimulus. In my own view the archetype, in the human being at any rate, has both these characteristics. We are on the one hand *affected* by the archetype from without; and on the other have inner experience of and give *expression* to the archetype by making images of it – by incorporating it in our art, our social behaviours and other products. On the one hand, then, we make the motor-car what it is; but having (deliberately) made it, we respond to – derive the instinctual 'benefits' from – the attributes which we have put there in the first place, but which still act on us very much as would, and does, any naturally occurring archetype or archetypal situation. This process of self-stimulation, incidentally, is also demonstrated, though of course less ingeniously, by animals – for instance when a cat pats a ball to make it roll and then chases after it.[8]

We give a car a fast reaction-time (acceleration) because, I suggest, fast reactions 'turn us on'. Why do we, as well, make the car shiny and hard?

This whole matter provides an excellent instance of a biologically determined and inherited psychological characteristic in the human being. No amount of practice raises (and I suggest no style of early environment would raise) the average female reaction time to that of the average male. That this innate difference may be one item in a general infrastructure of other such innate differences, producing or leading to *unalterable differences* in ability or performance at the intellectual and cognitive level, is a very clear possibility – I think, a probability.

[8] Movement is a releaser for both instinctive chasing and instinctive biting in many species of animals.

Taking these last two points in reverse order, the hardness is the hardness of well-toned muscle. The male body and its muscles in, or after, a period of active use are, literally, hard. That this is a generally desired condition is obvious from the sporting spare-time activities of males, particularly young males, and from the contempt with which we use words like 'flab', 'flabby', 'soft',[9] 'softie', 'paunch', 'gut', and so on, nicknames like 'fatty'; and from the very bad time that is often given to fat boys at school. During the moment of actual use also muscles become temporarily even harder, through the action of various autonomic processes, of which the hardening of the penis is an instance and special case. On the metaphorical level the phrase 'a hard man' is actually a tribute (though we are sometimes reluctant to admit this) in business, in cowboy and gangster novels, and so forth, and as witness the actual nature of many of our screen idols.

The *shininess* of the motor-car, however, is a female attribute.

Let us approach this matter somewhat obliquely. It seems to us so self-evident and 'natural' that shininess is a desirable characteristic that it perhaps surprises us to be asked why this should be preferred to a dull or matt finish. Yet, from a utilitarian point of view, a dull or matt finish (particularly in cars) is better – it reveals splashes and marks far less, hence requires less cleaning and upkeep. And what too of the considerable amounts of time and money which the owner habitually and willingly expends maintaining and *further improving* the shine on his car? What, too, of the love of the large amounts of functionally useless bright-work – of chrome flashes, strips, caps, and so on?

We note that the female in general has far less body-hair than the male. This seems to be a firm evolutionary trend. We note further that, at least in the West, woman goes to great lengths to remove what body-hair she has (and also that many men find facial and chest hair in a woman actually disgusting). We note also that a woman's skin is smoother than a man's. We note that lipstick in general makes the lips shine (matt lipsticks do exist, but they are not dull-matt) as well as making the lips more slippery, and that a woman will often wet her lips with her tongue before kissing. This again causes the lips to shine and become slippery. We note again that silk and nylon stockings and underwear are shinier even than the smoothest skin, and that their sheen is increased when the material is tight. Thus stockings must never be wrinkled – their appeal is in their sheerness. A tight dress is one sure-fire method of arousing 'wolf-whistles' (the more precise outlining of the female

---

[9] 'Soft', by contrast, is one of the nicest attributes when applied to women.

shape being, certainly, an added bonus). And we note, last and not least, that the fetishist adores tight rubber and leather clothing on women, highly polished shoes, shining head hair, and so forth. In sum, the male is quite certainly 'turned on' by the shiny and smooth.[10]

When he gets into his car, apart from extending and exaggerating his Ego, the driver is also getting into a woman.[11] The car is habitually referred to as 'she'. The driver, in a sense not too far removed from sexual reality, when driving is 'having' his car. Conversely, the woman passenger, particularly in a sports car, is being 'had', a view that women, as far as I know, have not quarrelled with.

The archetypal underpinnings of business and commercial life have begun to attract the increasing attention of writers in recent years, and there is growing awareness that no mere metaphor is involved in the parallels between hunting, warfare and business practice – so that, for instance, the jargon of the commercial world is littered with battle terms. I would like to take the general notion somewhat further. I propose, specifically, that the profit on a transaction is to be regarded as the symbolic kill, and that what is being satisfied in the individual concerned is a very strong archetypal urge both to kill and to be seen to kill. In the total situation we are considering, this urge is of course also integrated with and buttressed by the target archetype, the hunting archetype (one we have so far not discussed, which survives very much in the actual fox-hunt, grouse-shoot or whatever, and again in all kinds of games), and many others.

I do not wish to open up a full discussion of the complex mechanics of capitalism and economics, which would be necessary were I to attempt to meet all probable objections to my proposal. Instead I would simply point to the central role of the profit motive in all forms of commercial transaction, and in particular in the balance sheets of public companies. There are many, many criteria on which we *could* judge a company's value – in all senses of that word – but the world at large nevertheless chooses as its criterion the size of profits : how large a profit, how quick a profit, how much bigger a profit than that of other companies. Whether the product involved be plastic seaside souvenirs, sweets, cigarettes, rifles, inflammable plastic jelly for killing human beings (or, conversely, medical sup-

---

[10] What we see in the foregoing is an instance of man 'bettering' nature. This is a crucially important behaviour, and the clue, I believe, to many of the mysteries of evolution. This will be among the subject matter of later chapters.
[11] This double function of the one symbol of course presents no problem in symbolic terms.

plies, educational aids, and so on) is, it seems, a matter of no real relevance or importance. Other criteria we could use instead to assess the 'worthwhileness' of a particular industry or company would include the beneficial or detrimental effects of the work on the well-being, both physical and mental, of its employees, the waste or conservation of natural resources, and so on. But none of these other criteria come into consideration – *except* of course in the dehumanized sense of whether any such aspect influences the one essential point of the exercise – *profit*.

I am aware that a business man would point out impatiently that in the foregoing I am failing to consider money as money – and that economics demands that one put this aspect above and apart from all others, no matter what one's feelings. As I have said already, I have not the space here to consider the nature of economics and economic laws. But perhaps one small attack at the strictly financial level is in order.

I asked an economist friend whether he knew of any accounting system that considered the wages of the employees as profit. He said that as far as he knew wages were always considered a cost, and profit was what remained after all costs, including wages, were met. It seems clear, then, from this and our earlier comments that 'profit' is regarded as something you (that is, the investor, capitalist, entrepreneur) can take back, can *take home,* once the whole operation is completed. This does, of course, seem self-evident in the very meaning of the word profit. And yet, if after your undertaking you have covered all costs (raw materials, capital equipment depreciation, and so on) *and then still have money to pay wages*, is *this* not a profit? You have, after all, 'made' money that was not there before. That is, you began, say, with a capital loan of £1,000. You ended with £1,250. After paying back the loan of £1,000, plus any interest, you have left (say) £200. The fact that this sum is required for, and is distributed as, wages and does not go into your personal pocket, nor appear in the appropriate column of the balance sheet, does not stop it being a profit. Yet apparently neither the business man nor the accountant is prepared to regard this as a profit.

It appears that the essence of profit, to say this once again, is that which the entrepreneur takes home afterwards. Does this definition of profit suggest, then, that what we are talking about after all is the 'kill' which the hunter brings back from the hunt, among other things to prove how good a hunter he is?

There are, I consider, also *archestructural* levels concerned in the capitalist enterprise. The employer or boss is Ego; the worker is Self. It is in the essential nature of the Ego to take from, not give to (the Self). Wages (to workers) must and will therefore be kept as small as possible – only the amount, perhaps, that will keep the worker fit enough to go on doing what he is doing, but no more.

The idea of profit-sharing with other Egos is bad enough; that of distributing profits to workers seems anathema.

Conversely, it is in the essential nature of the Self to give, not take from (the Ego). This is the other half of the otherwise surprisingly enduring equation. The Self – as worker, woman, and so on – is predisposed *to accept domination and exploitation as proper*, as being the Ego's just due. This is one reason, I suggest, (though only one) why the oppressed, of all kinds, tolerate their chains, of all kinds, beyond, and sometimes far beyond, a reasonable point. In some sense of that expression they do 'love their chains'; and here we see a link with the masochism discussed in Chapter 1. In psychological/biological terms all these individuals are caught in an archestructural situation that reduces, and sometimes very heavily reduces, their freedom of action as whole, independent personalities.

I have deliberately passed rather quickly through some of the foregoing rather controversial possibilities in order not to sidetrack too much from the main drift of the chapter. Our course has brought us, designedly, to a consideration of the broader aspects of Ego.

(ii) Ego as Organism

Regarding the non-dual archestructure, there are in all four theoretical positions or possibilities: the Self's view of Self; the Self's view of Ego; the Ego's view of Self; and the Ego's view of Ego. One instance of each of the *first three* possibilities is, respectively: the moon; the sun; the witch.

It is the fourth possibility, the Ego's view of Ego, we have principally in mind when we speak of the Ego-archestructure.

There are clear and opposite differences between Self- and Ego-archestructures. The further back in time we go, for example, the more we find of the *former*; the further forward we travel the rarer, and weaker, these become. The opposite holds for the latter. With each step forward in time the power and province of the Ego-archestructure increases.

The Ego-archestructure, in sum, is the machine. 'Machine' in the present context refers to any man-made, functional object or artifact. The first stone picked up and *used* as a weapon by proto-man contained the seeds of the first Ego-archestructure; as perhaps the first Self-archestructure was a twisted tree, or the wind sighing through a crevice. The latest, more powerful Ego-archestructure, and the one we shall be considering in detail, is the computer.

To jump to our end-point, it will be our task, for reasons which will become clearer, to demonstrate not only that the computer does not, and cannot in any sense of the term, think – still less possess, or possess any possibility whatsoever of possessing, any form of consciousness – but also, contrary to repute, does not even and cannot : select, store, retrieve, scan, analyse, inspect – or whatever. To show that only organisms, and principally the human organism, can perform such actions.

A case study by the psychiatrist Bruno Bettelheim – 'Joey: A "Mechanical Boy" '[12] – is of great interest and relevance to our discussion. This boy, a victim of childhood psychosis, believed himself to be a machine.[13] He had attached to himself all manner of wires, vacuum tubes and similar paraphernalia, as also to his bed. He believed that these contrivances powered and were responsible for his being, and that without them he would die. He switched himself on in the morning and off again at night; and he assigned special functions and powers to certain parts of his equipment. Thus, unless a particular piece of apparatus was 'functioning' when he went to the toilet, he believed that he would be sucked down by the toilet and destroyed. He frequently made statements to the effect that 'machines are better than the body.' They were, he said, hard, could not be destroyed, did not feel pain.

As Bettelheim shows, Joey's psychosis was an attempt – of course a doomed attempt – to deal with his unbearable human and emotional problems. Only in the course of many years of institutional care was the boy in time able gradually to dismantle and dispose of his machines, and at the same time to begin to verbalize and confess to some of his terrors – for example, his belief that the world was a sea of excrement.

Joey's case is a very extreme form of what starts out as – and *is* – a perfectly reasonable process. In the course of his early evolution man realized, or discovered first by accident and then realized, that, for instance, objects hot from the fire could be manipulated by pieces of wood or slivers of stone; that hitting another individual with your hand was less effective – and more painful and personally dangerous – than hitting him with an object; that remembering how many days you had travelled was more reliably effected and recorded by putting a stone into a skin pouch every morning, instead of attempting merely to remember the number of days; that cutting a mark on a tree could pass information to a second individual, even though you yourself were not there in person, and so on.

These kinds of discoveries – action at a distance, action by proxy, the recording of events other than by means of memory – were and are, in a very real sense, the foundations of our civilization. The

---

[12] *Scientific American* (March, 1959).
[13] Only perhaps the intensity, but not the content, of Joey's delusions is unusual.

way of life and the standard of life we enjoy would be unthinkable without them and their sophisticated descendants. These processes, let it be clear, are in themselves wholly admirable and wholly desirable, and we – man – owe System A and the Ego an incalculable debt in this general context. Unfortunately, this is not the end of the matter.

For evolutionary reasons on which I hope to be able to shed at least some oblique light later, the Ego's appetites are frequently insatiable. The Ego is not content to let Caesar's laws operate in Caesar's domains, where they are apposite. It is not content to control external, objective reality. Instead it seeks to control, and to evaluate, by these same otherwise admirable means, the inner, subjective world. It does this partly because it is in the Ego's nature to regard everything outside itself as a potential enemy, and anything outside its control as potentially dangerous. *That* situation, while not ideal, is at least possible – that is, workable. But the Ego does not even remain – perhaps cannot be – content merely to supervise, or merely to monitor, that which lies outside its (true) sphere. It seeks finally to eradicate that which is not itself. At this point the rational tips over into the unreasonable. In the ultimate analysis, no longer content to direct, the Ego, having grown to believe it *can* do so, now decides that it *must* destroy, once and for all, its ancient enemy, the Self. With the Self out of the way, all anomalies will have been dealt with – the very source of anomaly. This is the 'final solution'. With this step man (meaning the Ego) will achieve at a stroke (a typical Ego metaphor, the killing with a weapon) its ultimate goal : will have conquered pain, disappointment, inefficiency, weakness, superstition, mediaevalism, filth – all these unfortunate legacies of an 'animal' past.

The only difficulty in this plan is that the despised, feared Self is not outside us, but only 'outside' the Ego. We are applying the metaphorical surgeon's knife to our own being. And in that process we destroy the very thing we seek to save. Joey is a terrible instance of this. I think also of the surgeon excising sections of the brain; or of that psychiatric illness where the patient cuts pieces from his own body.

The computer has many archetypal features – speed of reaction time and literal hardness, for example. The hero archetype is also present (metaphorically rather than literally, perhaps) : for one thing, it has the 'best' memory. More significantly it is commonly regarded as that instrument which will conquer for us the remainder of the universe (though I think it is more likely to destroy it). It is the hero, the super-machine, the unbeatable champion. Yet, all this is still the least of the matter.

For in the computer the Ego-archestructure has reached its full, and actually its first true flowering. It has reached the culminating point from which the Self-archestructure started out (as usual a reversal of situation). That is, in short, *the computer is felt to have life.*

This important difference between the two types of archestructure remains – that the Self finds its archestructures in, and injects its life into, natural objects (a tree, a mountain, the sun); while the Ego finds its archestructures in and injects life only into what it has made itself.[14]

The horror of this situation – this point in time at which the Ego of man trembles on the brink of psychosis – cannot, in my opinion, be exaggerated. The problem is how to persuade mankind of the true nature of the situation, while it may yet be reversed. How and where may one begin? Perhaps as follows.

If I hit a metal piping with some other object in such a way as to make a mark on its surface, would one say that the piping has 'noted' or 'recorded' the action? No, one would not. At this level, even the purely metaphorical application of such terms feels inappropriate. We would say instead that the piping 'is marked' or 'has a dent in it' – the statement quite appropriately incorporating the passive, inert nature of metal piping.

Similarly if I threw first a ball of cotton-wool, then a cannonball, at a sheet of ordinary glass, with the result in the second case that the glass shattered, I would not say that the pane of glass had 'sorted' the two objects into two categories. And if I were to say this, another person might well look at me in some puzzlement, wondering even if I were altogether right in the head.

To take one final example, would it be a reasonable use of language, let alone conceptually acceptable, if, after rolling a marble towards a hole in the middle of a table, so that it fell through, I said afterwards that the marble had chosen to fall through? No, indeed, it would not.

Imagine now for a moment a series of inclined, interconnecting metal gulleys arranged in structural form, having in them at various points holes of a size through which a marble could fall. And imagine, further, that the gulleys are somehow pivoted, so that pressure at one end will reverse the angle of inclination. And let us suppose that I set marbles rolling at various points in this structure. There will be for a few moments a kind of aerial ballet – perhaps even an aesthetically pleasing one – of marbles rolling and dropping, of gulleys tilting and retilting. But we will be clear, I hope, that any literal application of any such words as 'decision', 'choice' and 'res-

---

[14] The reverse side of the same coin, of course, is the Ego's *denial* of life (consciousness) to the organic.

ponse' to describe the movements of the gulleys and marbles would be quite misguided.

Let us turn to a car engine. This is a quite complex and complicated machine. In addition, it has certain features which resemble – or possibly mimic – features of our own physical being – diaphragms, valves, 'intestines', 'veins', and so on (and perhaps by no means altogether necessarily, two 'eyes' and four 'legs'). In these circumstances it is perhaps understandable that we do not always regard the car engine simply as a mechanical contrivance, but rather as an organism or person. We say of the engine, and of cars as a whole, 'it's a bit temperamental', 'it goes when it feels like it', 'it's acting up this morning', 'she's a reliable old thing', 'the damn thing *knows* when I've got an urgent appointment', and so on.

But still and all, if one were to press drivers, and ask whether they thought the car *really* had any feelings or any awareness of any kind whatsoever, the vast majority would reply, I am sure (albeit perhaps in one or two cases reluctantly), that, no, a car does not and cannot possess any such qualities or capacities. A car is only a machine. The feelings we speak of arise within ourselves, and are projected on to the car.

What, finally now, of an adding-machine? This item is of considerable interest to our discussion, because it is actually very like the computer in some respects, and certainly much more so than a car engine. A modern adding-machine, moreover, as the salesman or instructor observes, not only adds and subtracts, but also divides, multiplies and calculates squares, sums of squares, square roots. (Actually, of course, the machine does *none* of these things.)

Matters are best understood if we consider the hand-operated variety. In that machine, when one turns a handle through one revolution, one or more cogs or levers catch in one or more other cogs or levers causing these others to move – or, of course, alternatively to have their own motion arrested by them. The various cogs and levers are in turn and finally connected to 'faces' which have the digits 1 to 9 variously marked upon them. These show at and are removed from panels at the top of the machine, whence we read off the 'answers' to the 'calculations'.

The machine does not 'add up' or 'subtract' numbers because it is quite incapable of either of those functions. Only human beings can add and subtract. The concepts do not exist, outside of our heads. Cogs move cogs, or not, levers move levers, or not, in accordance with the physical laws of the physical universe and for no other reason (that is, because of pressure and counter-pressure), and never with any deviation from those laws except as provided for by those laws. An unnecessary rider this, of course, because in nature *all* laws are laws and exceptions are only further laws: the 'exception' is itself once again a wholly and totally human concept that exists nowhere else in nature.

Still less does this adding-machine of ours divide and multiply – this time not even in *any* literal sense, leave alone conceptually. When the machine is 'asked' by us to divide, what it does is to 'subtract' the divisor repeatedly (and very quickly) from the dividend until nothing or a remainder is left. In so doing it 'records' the number of times it has performed the operation and, as said, 'leaves' any remainder. The mathematical notions of dividing – either of (a) conceptualizing the possibility or of (b) performing an operation to find out how many times one number is contained within another – are wholly, and solely, human ideas. They do not exist anywhere in the universe outside the human mind. It is very necessary to be repetitive on this point.

The foregoing remarks apply equally to the fully electrically operated adding-machine. Instead of our hand turning a cog or moving a lever, this is done by an electric impulse; and of course the lever or cog is itself replaced by a further electrical impulse, some exchange of potential, or whatever. A sophisticated device of this kind can be 'instructed', i.e. so built that initial actions of ours – programming, essentially – cause certain possibilities to be as it were 'evaluated' by the machine, and then 'carried out' or not.

I have placed a number of the preceding words in quotes to bring out the fact that their use is purely metaphorical. A pane of glass cannot sort objects; a marble cannot choose to do something; a pivoted piece of metal cannot make decisions. Nor can a car be stubborn, for it is only a machine. No more than an adding-machine can add. They are all only machines.

The computer, too, is only a machine.

Thus it cannot store, retrieve, scan, analyse, process, choose, compare – or perform, literally, any human action of this or any other kind; any of the human actions to which, and only to which, those terms refer.

The second kind of adding-machine I mentioned is, indeed, a basic computer. To 'compute', in any case, is only a smart word for to add. If indeed we call the computer an adding-machine, that being all that it in fact is – though additionally equipped also (by us) to 'translate' its computations into whatever kind of 'actions' or 'information' we had in mind when we designed it – a great deal of the nonsense is at once thereby knocked out of the concept. The fact that it is an extremely complex and very cleverly designed adding-machine makes no difference, and is of no consequence whatsoever. Just as a simple car engine is not an organism, so a very, very complex car engine is no jot nearer that impossible possibility. No amount of complexity can bridge this difference of kind.

To be absolutely blunt about it, there is no essential difference between a room full of computer equipment and a room full of cups and saucers. Such differences as one might describe are differences only of degree. And in any case, only *we* know that. Neither the tea-cups nor the computer know this. Neither of them can ever know anything.

When one says that the difference between a tea-cup and an organism, or between a computer and an organism, is one of kind, what does one precisely mean? It is, of course, true that human thought, and the neural behaviour of all organisms, is accompanied by, or associated with, some kind of electrical discharge – more exactly with exchanges of electro-chemical potential at the molecular level. It is this fact (this wholly superficial resemblance of detail) which allows the Ego (the scientist) to jump, or ignore, the unbridgeable gap between organism and object – i.e. computer. The computer, he claims, is functioning just like an organism. Ergo, it is an organism.

Now, even at the most basic level it must be very seriously questioned whether a series of blips on a magnetic tape actually at all, let alone closely, resembles what goes on within the organism at the molecular level, for example when thinking is taking place. But let us be very, very generous and assume for the sake of argument that these two sets of processes are in fact identical or quite similar. Yet even then what a travesty of argument still remains.

If I produce by some means a noise identical to that of a car engine, such that anyone listening to it is convinced that he hears a car, have I thereby made or even simulated a complete car? Or (perhaps a better example) if by dissembling I convince a girl that I am in love with her, so much so that she would wager even her life on the truth of it, am I, then, really in love with her?

In these two examples is it not the case *that there are very large areas of difference not on view* between the real and the simulated phenomenon? This is how it is also in comparing the computer and the organism. The absolutely crucial difference between the two resides in the fact that in the human being (and more rudimentarily in all organisms, as I personally believe) the outcome of the operations at the molecular and para-molecular levels, at least potentially, *passes into awareness*. We become, and are, conscious of mental events. We *know*, and we know that we know.

The computer for its part has no awareness. No awareness of any kind whatsoever. It is not animated by any consciousness or life. It is *inanimate*. It does not know, or know that it knows, nor does it know that it does not know. For not knowing, in the sense which I am using the term here, is but a further form of knowing.

What precisely, then, the sceptic asks at this point, is consciousness or awareness – especially if you insist it is not the electrical activity of the brain? This is a question we shall be trying to answer in a later chapter – so that for the moment the query is merely noted.

I would now instead like to bring us back to our starting point – the computer as Ego-archestructure. What on earth *is* this fixed desire on the part of the Ego to create a living being which is not an organism – and to believe, in the face of all reasonableness, on the contrary that a computer is alive, or at least is on the very verge of so being? I must ascribe the phenomenon on the one hand to the basic split that only too readily occurs between the Self and the Ego in the formative years, a split that once set under way, be it by predisposition, early environment, a chemical imbalance or whatever, frequently seems to feed upon itself and become ever wider. I must possibly ascribe it, too, to evolutionary error on the part of nature. It may be that in the Ego we see an irretrievable specialism or specialization of function: an adaptation carried so far that it cannot further adapt (that is, differently adapt) to altered circumstances – or even to the circumstances that produced it in the first place. This last remark is a trifle obscure – but partly I am concerned not to open up a further area of discussion for the moment. I ask the reader to bear in mind the possibility of a solution that in the end produces out of itself difficulties that finally defeat the solution – as when in a crossword puzzle a wrongly solved clue can carry us forward to (temporary) further success, but ultimate disaster. So the following remark of Tinbergen: 'An increasing, but still far too small number of people begin to realise that we are caught in a vicious circle: the very success of our behaviour has led to a situation from which only a better understanding and controlled change of our behaviour can extract us.'[15]

Be these possibilities as they may, it of a surety remains true that the Ego desires nothing more avidly than to creep away into some machine, and to leave its feelings and its body and its essential humanity to their fate. So strong is this desire – which I think to be totally mad – that otherwise fine intellects such as Arthur C. Clarke apparently take this non-fact, this non-possible fact – the animate, living, conscious machine – to *be* a fact. No one, I think, would ascribe life to a hammer in his hand. Why then ascribe it to a heap of lifeless metal and plastic – even if it is conducting a low-level electrical charge? One might as well say that an electric fire or a radio is alive.

Like its Self-counterpart, the vampire, the computer exists – that is, has life – only if you believe it to have life.

---

[15] *The Study of Instinct* (O.U.P., London, 1951).

But for all its efforts, all its fantasies, we have (I trust) seen in this section that the Ego is after all, first and last, only an organism, and part of an organism. Even when it goes quite mad, as Joey did, it does not succeed in being anything else.

III

BIOLOGY AS DESTINY

3
Body-Types and Personality

The general title of this short section is 'Biology as Destiny'. The title could, however, have been applied equally to the first two chapters of the book. It is by no means incorrect to consider that the general argument of this book *first* goes out of its way to show man (and organisms in general) as nothing more than biological facts or artifacts. In this particular connection I have indeed elsewhere spoken of the personality as the genetic and the biological 'prisoner'. But later in the book I will nevertheless reverse this argument, in an entirely crucial sense. For if one speaks of a 'biological prisoner', does this not at least imply the imprisonment or incarceration of something or someone *other* than a biological event – and the possibility, too, of its release? A definition of these various terms, however, and the rather inflammatory issues that are thereby raised, is avoided for the moment.

Nonetheless, to revert to the specific purpose of this chapter, the currently somewhat despised contribution of body-type theories to our general understanding of man and his nature will appear, for the time being, to reinforce still further the (biological) notion of physiology as a primary, if not indeed the sole, determinant of human personality. Body-type theory – or constitutional psychology, as it is also called – is certainly the most extreme expression of 'biology as destiny' to be found in the psychological literature. In considering these theories I am concerned not so much with their merits and demerits *per se,* or with their individual internal ramifications and differences, but with substantive parallelism with, and support for, my own description of the human organism and its evolution. I shall be drawing principally on the views of the psychologist W. H. Sheldon, as set out in his *Varieties of Human Physique* and *Varieties of Human Temperament*[1] (with a glance, too, at Ernst Kretschmer's *Physique and Character*[2]) and C. S. Hall and G. Lindzey's *Theories of Personality*.[3] The reader is especially recommended to Chapter 9 of the last-mentioned book.

The attempt to relate temperament systematically to general body features is usually considered to have begun with Hippocrates. Broad non-systematic opinions no doubt long preceded him, running as they do through the popular imagination of all ages up to and including the present day. Thus fat people are said to be jolly, thin people to be morose, red-haired people to have quick tempers, and so on.

Hippocrates proposed a two-part classification, dividing mankind into those who were short and thick and those who were tall and thin. This proposition is of considerable interest and relevance in the light of our earlier discussion of the comic and tragic personifications of Greek drama – and their suggested respective relation to Neanderthal and Cro-Magnon.[4] We shall not lose sight of Hippocrates' basic pair in going on to consider the format of numerous body-type theories since his time, comprised as these later ones are of principally *three* basic types.

Precise details of the numerous constitutional theories devised and put forward over the last two thousand years or so do not concern us. Sheldon lists some twenty-eight latter-day theories, emphasizing that this is a small sample only. We are here considering but two, and these only briefly, principally Sheldon's own and that of Kretschmer, a German psychiatrist active in the first quarter of this century. Perhaps the essential point for us is the large areas of overlap each of these many theories shows with the others. While the protagonists themselves no doubt saw many differences of detail and emphasis between their own and others' ideas, to the outsider the impression is of the same basic facts being worked and reworked over and over again. And, indeed, Sheldon himself does not seem to disagree with this view.

As stated, the vast majority of the theories, with the exception of Hippocrates', are tri-partite. A few sub-divide one of the categories (the third) into two types or sub-types – a variation not of great significance, except I think from one point of view, which I shall be indicating.

---

[1] Harper, New York, 1940, and Harper, New York, 1942.
[2] Springer, Berlin, 1921; Harcourt, New York, 1925.
[3] Wiley, London, 1957.
[4] Many later studies of the twentieth century, incidentally, have shown low but persistent correlations between mental ability and the tall, slender physique – a further link with, and prop to, my general arguments.

Kretschmer names his three types the *pyknic*, the *athletic* and the *asthenic*. The first of these, the pyknic, is characterized by plumpness and roundness, has 'a soft, broad face, a short massive neck sitting between the shoulders . . . a magnificent fat paunch protruding from the deep vaulted chest'. The second is a muscular, vigorous physique 'middle-sized to tall', has 'particularly wide projecting shoulders, a superb chest and firm stomach'; the pelvis, arms and legs, though strongly formed, are 'almost graceful' by comparison. The third variety, the asthenic, is a frail, linear physique. It has 'a deficiency in thickness' in all parts of the body 'combined with an average, unlessened length, a skin poor in secretion and blood . . . and narrow shoulders from which hang lean arms with thin muscles'. A small group having a *markedly* divergent mixture of these otherwise in general linked characteristics, Kretschmer termed *dysplastic* – a group to which, again, I shall refer later. Perhaps it should be pointed out respecting all constitutional theories that individuals as we find them in daily life rarely if ever correspond exactly to the theoretical 'pure' type. As many theorists freely admit, most individuals possess or comprise a mixture of attributes. It is a question rather of which tendency *predominates*. On the basis of this predominance the individual is assigned to one of the major groupings. The dysplastics mentioned are essentially those impossible to categorize even on this basis.

One of the most interesting of Kretschmer's findings was a particular relationship of individuals of the body-types described with two major forms of psychosis – namely manic-depressive and schizophrenic psychosis. Examining a total of 260 psychiatric patients and sorting them into his categories, Kretschmer found the following relationship with the two forms of psychosis named:

Table 1.

	Manic-Depressive	Schizophrenic
Pyknic	58	2
Athletic	3	31
Asthenic	4	81

This result firmly associates manic-depressive psychosis with pyknics, and shares schizophrenia between the athletics and the asthenics, with a clear bias towards the asthenic.[5] Is not the first of

---

[5] With regard to these results, as also those of Sheldon and other theorists, independent investigators using the prescribed methods of classification, and so forth, rarely produce as convincing or clear-cut a result as the experimenter himself. Nonetheless, a fairly high order of confirmation is usually obtained, and the findings I report here and elsewhere are not in dispute, as far as their local terms of reference are concerned.

these two findings of particular interest in connection with our discussion of the nature of the B-dominant in Chapter 1? The link shown by Kretschmer between athletics and asthenics is equally useful. For, not to beat about the bush, I intend to propose that the three types of this and other constitutional theories are none other, in broad outline, than the embodiments of my own three personality divisions B1, B2 and A. I have suggested frequently that the A-dominant evolved from an earlier B2-dominant. The link between the two corresponding types of Kretschmer's theory (athletics and asthenics) is, then, both expected and welcome. Nor, of course, am I at all sorry to see a relatively greater incidence of schizophrenia among A-dominants – Kretschmer's asthenics.

I do not intend to examine any more of Kretschmer's views and findings, since the remaining points of interest are rather better brought out for our present purposes by Sheldon's results.

Sheldon's three types are referred to as the *endomorph*, the *mesomorph*, and the *ectomorph*. The general appearance of these three types does not differ significantly from the descriptions given by Kretschmer, so that I will not elaborate on this aspect of the matter further. The terms Sheldon employs for his types are, however, derived from embryology. For Sheldon claims that the dominant organs of each type (in the endomorph the digestive viscera, for example) derive from the endodermal, mesodermal and ectodermal embryonic layers respectively. Now, as it happens these are the inner, the middle and the outer layers of embryonic germinal tissue. I take the case to be that the innermost layer is the *oldest* and the outermost the *youngest* in evolutionary terms. How interesting a 'coincidence' that the rank order in evolutionary age of Sheldon's three types corresponds precisely to the rank order in evolutionary age of my own B1, B2 and A types. How interesting too that the endomorph – type B1 – derives from the *innermost* layer.

There are numerous scattered observations in Sheldon's work that I find supportive for my own views, and a minority that are in rather direct opposition. For example, the skin of the male endomorph is described as 'soft, smooth and velvety' and the pubic hair shows 'the so-called feminine pattern'. Yet 'a premature tendency to baldness is often seen, even in youth'. The first two comments are entirely acceptable, the last not at all so – for of course *women* seldom go bald at any age. Why should we find baldness, let alone premature baldness, in the 'female' (B1) male? I have no ready answer for this, other than to see it as an unlooked-for side effect of hybridism (a not unprecedented event), man being, as I have elsewhere argued, very much a hybrid species. My general theoretical position is notably saved, however, by Sheldon's findings that *females* (the nearest we get, of course, to a true B1-dominant) are as a whole much more endomorphic than males.[6]

With the various physical types and methods of classification, measurement and so on described to his satisfaction, Sheldon and his co-workers turned to the behavioural spectrum. Combing the psychological literature, they produced a more or less exhaustive list of temperamental and attitudinal characteristics. These they were able to reduce firstly by general agreement amongst themselves to more manageable proportions, through eliminating trivial items and by combining into one term items evidently overlapping. This list of attributes was then applied to a sample of graduates and undergraduates with a view to establishing which, if any, of these psychological characteristics were correlated. Three clusters or groupings of attributes were observed. Following this preliminary work, a good deal of further research and refinement led ultimately to Sheldon's Scale of Temperament – three groupings of twenty characteristics each. These three component-groups respectively describe or comprise *viscerotonia, somatotonia* and *cerebretonia*. The three temperamental types yield high positive correlations with the three types of the physical scale – endomorphy, mesomorphy and ectomorphy.

The majority of Sheldon's behavioural or temperamental traits call for no special comment in relation to my own general theory, but a few are of great interest. Thus associated with viscerotonia (endomorphs) are: slow reactions (compare here our discussion of female and male reaction time in Chapter 2); orientation to people; greed for affection and approval; deep sleep; relaxation and sociophilia under alcohol and orientation towards childhood and family relationships. One need hardly point out that these are characteristics representative of *woman* in general (e.g. deep sleep). Associated with somatotonia (mesomorphs) we find: competitive aggressiveness; spartan indifference to pain; assertiveness and aggression under alcohol and orientation towards goals and activities of youth. How well these reflect the B2 'male' personality. Alcohol here has the effect of releasing the aggressive ego-instincts – as contrasted with the sociophilia (the release of libido-instincts?) in the B1 personality. The third component, cerebrotonia (ectomorphs), is associated with: overly fast reactions; socio*phobia*; poor sleep habits; youthful appearance; resistance to alcohol and orientation towards the later period of life. Here we see, I suggest, principally in the avoidance of other people and the flight from the whole of the un-

---

[6] While adequate samples had then not yet been tested – hence I do not wish to stress the point overmuch – it appears that mesomorphy is more common among African Negroes than either of the other two components. This agrees well with my own view of the Negro as a B2 dominant.

conscious (poor sleep, resistance to alcohol), the ground-plan for the schizophrenic withdrawal that this personality undergoes in breakdown.

I was especially struck with the last trait in each of these three series – orientation towards childhood, orientation towards youth and orientation towards the later period of life. How well this mirrors the derivation of these behaviours from the endodermal, mesodermal and ectodermal layers of the embryo : the most ancient (i.e. the earliest in evolutionary terms), the next most ancient and the least ancient (i.e. the latest) germinal layers. How well too it reflects my own evolutionary sequence of B1, B2 and A. As I argued at some length in *Total Man*, the child is a B-dominant (the very young child a B1-dominant), the adult an A-dominant.

Several testable hypotheses arise from this comparison of my own and Sheldon's views. It should be the case, for instance, that endomorphs (or pyknics) score higher on tests of extrasensory perception than ectomorphs (asthenics), More spiritualist mediums, therefore, should be endomorphs than ectomorphs. Endomorphs should also dream more – a hypothesis already half borne out by the fact that they sleep more. They should also hypnotize more easily than ectomorphs, and so on.

There is the matter of dysplasia. Both Kretschmer and Sheldon use this term and have this minor category. It is composed of those individuals who exhibit such extreme mixtures of the basic components of the three types that to place them in any of those categories would involve undue distortion. Sheldon reports that dysplasia is more associated with ectomorphy than with either of the other categories. In this connection one notes too that, with those theorists who have produced not three, but four types, the group which all sub-divide is the ectomorph-equivalents, not either of the others. Finally, Sheldon reports ectomorphs to be the most neotonous group – the most 'youthful' in appearance.

All this suggests to me that ectomorphs *are still in the process of evolving as a variety*. Darwin states that the emergence of a new variety is accompanied by a general increase in variability in the evolvent population. A burst of variability, throwing up a large number of modifications, would of course add to the chances of the emergence of a favourable adaptation or adaptations, leading ultimately to a new variety or species.[7]

---

[7] The relatively high incidence of dysplasia noted in women I take, however, to have a quite different origin – namely in the 'domestication' of women. As Darwin showed, variation increases under domestication – that is, under protection, and removal from the evolutionary 'front-line'. I shall be defining what I mean here more carefully in a later chapter. Briefly,

Finally, let us consider the relations which Sheldon found between his three types and psychosis. His *three* (not two) categories of psychosis are affective (i.e. manic-depressive), paranoid and heboid (i.e. schizophrenic). Giving these conditions their more familiar names, Table 2 shows the correlations obtained by Sheldon between constitutional type and the form of illness.

Table 2.

	Manic-Depressive	Paranoid	Schizophrenic
Endomorphy	+.54	−.04	−.25
Mesomorphy	+.41	+.57	−.68
Ectomorphy	−.59	−.34	+.64

Acceptable features of this table are the involvement of mesomorphs in manic-depressive illness (though the correlation is higher than one would like), for B2-dominants are closer to the B1 types (are more 'emotional') than are A-dominants; as is the negative relation of the last mentioned, the A-dominants, to the illness. The high correlation between ectomorphy and schizophrenia is likewise expected. What is not expected is the greater negative correlation of mesomorphy than endomorphy with schizophrenia (namely −.68 compared with −.25). But, as it happens, three out of the nine correlations were not expected by Sheldon either! That is to say, my theories are no more (though also no less) threatened by these unexpected findings than are Sheldon's own.

The positive correlations between mesomorphy and ectomorphy, and paranoia and schizophrenia, respectively, are welcome and explicable. For my own view of all psychosis is that it involves a flight from the demands or the presence of the Self. Paranoid psychosis would be a fear of the Self, but with the anger and a tendency to defend/retaliate of the B2 personality in evidence. Schizophrenia would be the total abandonment of any attempt by the A personality to stand its ground or to face what should be faced – representing a condition of total flight from the Self. Still in these terms, manic-depressive psychosis for its part shows a would-be free Ego permanently bound to the Self in such a way that all attempts to escape (that is, the manic phase) merely as it were stretch the bonds, resulting in continual, subsequent recoil back into the Self (the depression). Thus expressed, manic-depressive psychosis would be the

---

whereas battle and hunting would as it were tend to *sharpen* the outlines of the type or types among males most likely to survive, those particular, quite lethal, pressures would not be working to anything like the same extent on the women at home.

closest of the psychoses to the other major group of mental illnesses, the neuroses – and if so, logically so, since commonest among endomorphs. Neurosis is, in the view I have expressed elsewhere, the virtual enslavement or capture of the Ego by the Self.

In conclusion I must emphasize that I do not regard the various trios of body-type theories of personality as being in anything like complete overlap with my own three types of B_1, B_2 and A. Nor do I necessarily accept any of Sheldon's detailed points of view, or any further ramifications of his theories other than those that I have discussed here. This rider applies equally to all other constitutional theorists and theories. I *do* very much suggest, however, that the evolutionary processes I myself envisage, and attempt to describe, underlie and are responsible for *both* the body-type theories *and* my own.

I consider also that the archestructural, or perhaps merely unconscious, perception of the three human varieties or sub-varieties which make up our hybrid species is one major factor that has helped produce the 'three' of folklore – three wishes, three Wise Men, 'accidents happen in threes', the eternal triangle, and so on. A further influence, however, has been, I think, the earth, moon and sun of our immediate macro-spatial environment. And finally, I have of course insisted that we are all archestructurally very aware of the Self, Ego and Person of the total mental landscape as I envisage it – a further highly influential trio therefore.

IV

THE DECLINE AND FALL OF BIOLOGY

4
The Energy of Evolution

The first time (I believe, in discussion with a lecturer in psychology, a good many years ago) that I heard it questioned that thought precedes language, and the reverse maintained, namely that thought proceeds *from* language, I think even then I almost wept. The reaction was something of the desolation felt when, in the course of some dispute, the testimony of an official that one knows to be a lie but has no way whatsoever of so proving, is accepted against one's own. How (I felt) could anyone question or deny this self-evident truth about thought and language – namely that language is the expression of an already accomplished process? It was the denial of the inner life, a negation I was to hear expressed many times in many forms from my psychologist colleagues in later years.

Before going on, perhaps I should say at this juncture that I do not at all dispute that language (be it the language of words or number), once evolved, is an aid, and even sometimes an essential aid, both to the refinement of thought and to the generation of further thought. This is a cycle, an as it were self-perpetuating dialogue between thought and language, which, once begun, continues throughout our lives. So effortless is this constant dialogue or cycle that it is not difficult to lose entirely the sense of dialogue; and to believe that thought is language, language thought.

But this 'harmless' view becomes an error of the greatest magnitude if we maintain it either absolutely or particularly in respect of the evolving or newly born human organism. Or, still worse, when we claim the *dialogue* to have *begun* with the word. My reasons for my views will become more apparent as this chapter proceeds. Apparently not more than a debating point, a mere theoretical difference perhaps between schools of thought, the dethronement of (inner) thought through the enthronement of (outer) language is, on the contrary, a significant step along the path which leads to the empty organism; the ultimate horror which quite eclipses the horror of all the machines-with-souls put together.

As I said a moment ago, for an adult to equate thought with language is an understandable error, if error at all. For as I suggest, so smoothly does the cycle, the dialogue, operate in later life that we no longer register it as such, and think instead of a single, unified operation. It happens, however, that thought without language is easily both demonstrated and induced. And since, perhaps rightly, an ounce of demonstration is worth a ton of faith, it is as well to demonstrate.

In the game of chess, a bishop moves diagonally, and a pawn moves vertically; but a knight moves — ? We all *know* how a knight moves in chess (like *that*); but our language has no term for it.

Children who have lost their parents are called orphans. Parents who have lost their children are called — ? Again the concept involved is clear. Again we lack any term for it.

The German word for friend is *Freund,* and the word for friendly is *freundlich.* The opposite of *Freund* is *Feind,* and means enemy. What does *feindlich* mean, in English?

The reader, unless he is unlike the dozens of students on whom I have performed this last little experiment, will have floundered from 1 to some 20 seconds, or even longer, looking for the word which names this perfectly clear idea (a word which in one of the more mysterious ways of the organism we are somehow aware *does* exist), namely 'hostile'.

Other very solid evidence of a different kind comes from the observation of young children in the process of learning to speak. Here we find that extremely complex conceptualization precedes the acquisition even of single words; and further that the words are assigned meanings and perform operations quite other than those of the 'real' word – those which the parent, of course, is very much trying to convey. Thus Jesperson[1] reports the instance of a boy (age given as 1.6 years) who, having once had a pig drawn and named for him, applied 'pig' to (a) actual pigs, (b) drawings of pigs and (c) *writing* in general. Another child, a little girl (between 1.6 and 2.0 years), used the word 'colour' for anything striking – an object, a situation – which caught her attention. She had learned the word in connection with a striking patch of colour in a picture. Another girl (at 1.3 years) used the word 'bang' for anything that dropped. At age 1.8 years she began consistently to use also a variant 'bing' for

---

[1] Otto Jesperson, *Language: Its Nature, Development and Origin* (Allen & Unwin, London, 1969).

(a) a door, (b) bricks and (c) building with bricks. It seems clear enough that she had taken or evolved the second word bing from the noise of falling bricks and closing doors; but was now using it as a noun – it is, perhaps, less clear whether she was also using bang as a noun – the name of a class of objects or situations. To meet her somehow altered feeling for this new category (either perhaps merely a new class of objects, or even a new grammatical or conceptual category, whichever was the case) she modified the vowel.

The next instance, which I myself recorded, is possibly still more illuminating. This concerned a little girl of rather less than eighteen months, the daughter of friends with whom I was staying. She had at that time a vocabulary of four words – 'mummy', 'daddy', 'doggy' and 'ducky', which she apparently used indiscriminately and without understanding. A few days of observation showed that she had in fact set up some very refined criteria indeed – constituting an almost global analysis of her non-static environment – which she applied with remarkable consistency. 'Mummy' was applied only to her mother and no one else. 'Daddy' comprised all human beings, including her father – that is, all men, women and other children. 'Doggy' (the house possessed a pet dog) referred to all living, non-human creatures – birds, cats, and so on – while 'ducky' was *all moving but inanimate objects* – trains, cars, boats, and so on. The last term, ducky, had derived from the bath situation, which she always took in company with several small plastic ducks, and without whose company she refused to be bathed at all.

The consistency and power of this child's observation were, as I say, astonishing. Thus the ducks in the park, despite their obvious resemblance to the ducks in the bath, were 'correctly' identified as doggy. And the object which another child pushed towards us – a life-size furry model of a dog mounted on wheels! – was firmly ducky.

Sandra's 'mistakes' were a constant source of amusement to her parents and their friends. They frequently and laughingly 'corrected' her. At these times Sandra wore a lost, puzzled, sometimes even a hurt expression; but she nevertheless stuck to her categories.

What we see here and in the earlier examples is an intelligence grappling – and grappling very well – with the complexities of the world around it. The process of concept-formation is already well advanced. *Only at this stage* does the further desire (?) arise to label and communicate the discoveries and realizations to others.[2] Though

---

[2] As a rule it happens that these are private, not publicly agreed, findings and hence early attempts to communicate them, as they are conceived, often fail. Only later does the child give up (most of) its private categories in the face of pressure from already existing public categories.

we need not necessarily urge a cause and effect situation, I do suggest that the urge to communicate both in the very young child and in prehistoric proto-man was and is a powerful motivating and causal force behind the evolution of language.

I hope it can be seen that a simple stimulus-response model is wholly inadequate when applied to these cases. Sandra's efforts in particular went entirely unrewarded, were unreinforced, by those around her. Indeed, she was mocked. Yet she stuck stubbornly to her guns. (Such 'reinforcement' as we may conceive of lay, I suggest, in the satisfaction Sandra felt in having made sense of her non-static environment.) Let us not overlook, too, the enormous possibilities for confusion which existed for her even in the *sounds* of doggy and ducky and which would tend to imperil conventional reinforcement processes.

Before my next comment I hasten to stress, in the interests of objectivity, that Sandra's concept-system was in full flower before my arrival on the scene (and, incidentally, that I did not come across Jesperson's chapter on this subject until some years later). However, I believe that after a few days Sandra realized that I understood her system. When she pointed to the dog on wheels in the park she looked at me, not at her parents, for confirmation, and made thereafter a point of categorizing our encounters for my special benefit. I did not detect one mistake.

My hope is that the whole of the foregoing has slightly predisposed the reader to think of the inner, experiential life as not simply some kind of useful adjunct to the understanding of behaviour, but a central and indeed prior issue in all behaviour, in the absence of which recognition all other studies are largely meaningless. To regard on the one hand empty machines as organisms, and on the other organisms as empty machines, is not simply an error. It is insanity.

I stress these points once more, for I wish to go on to discuss not merely the further subjective life of man, but, still worse, the subjective life of animals.

The current general anathema to subjectivity of any kind among academics and scientists concerned with behaviour is well expressed by Professor Tinbergen in the already mentioned book, *The Study of Instinct*. I shall be quoting his actual words on the matter. Nonetheless, Tinbergen's own further data are among those which we shall be using as the basis for a proposal which I think is a concept (and a subjective concept at that) of far-reaching consequence for our understanding of behaviour in general and of evolution in particular.

Professor Tinbergen writes as follows :

Because subjective phenomena cannot be observed subjectively in animals, it is idle either to claim or deny their existence. (p. 4)

Although, as we said before, the ethologist does not want to deny the possible existence of subjective phenomena in animals, he claims that it is futile to present them as causes, since they cannot be observed by scientific methods . . . Therefore, though I do not want to belittle the importance of a study of either the directiveness of behaviour or of the subjective phenomena accompanying our and possibly the animal's behaviour, I want to stress the paramount importance of recognising the limited nature of such a study . . . (p.5)

There we have it. A reasonably typical view of the phenomena of the inner life – if anything, one slightly more favourable than most. The subjective life of the animal, in particular, is 'only a possibility'. Is it not strange that scientists who are prepared to argue, and to try to demonstrate, that all other human behaviours and attributes are present in rudimentary form in lower animals – to claim very much that these behaviours have a long evolutionary history – are so reluctant to accord this same status to conscious awareness? Subjective experience, it seems, sprang into life unheralded and unprecedented with the first representative *sapiens*. I trust the irony of this last comment is apparent. The statement is patently absurd. As Darwin so often insisted, *natura non facit saltum* – nature does not progress by leaps.

Man himself has some kind of subjective life, Tinbergen concedes, and there is a possibility (only a possibility) of this in animals. Even so, he insists, such phenomena cannot be regarded as causal. He claims that he does not wish to belittle the study of these phenomena. Yet one's clear impression, nevertheless, is that he regards them as a by-product, and not more than a by-product.

I think our preceding examination of the behaviour of young children acquiring language, and the complex ideation which apparently precedes that stage, argues powerfully for a wish to express and communicate what is already experienced; and this is perhaps among the reasons why indeed a child learns a language.[3]

To particularize our argument now, it was discovered during investigations of the nature of sign-releasers by ethologists, that a wide range of organisms frequently respond more vigorously to releasers differing from those actually found in nature. These are termed 'super-normal' releasers. The naturally occurring stimulus situation,

---

[3] I do not, of course, wish to suggest that this is necessarily the sole explanation or that no maturational processes are involved.

then, is *not* the optimal one for the instinctive response. Tinbergen calls this finding 'remarkable', and indeed it is.[4] A number of instances are given in his book; general literature affords many others.

For example, the ringed plover, if presented with a choice between normal eggs (light brown with darker brown spots) and white eggs with black spots, prefers the latter. Oyster-catchers prefer a clutch of five eggs to the normal three. This same bird, if presented with its own egg, a second much larger, and yet a third enormous one, habitually responds most vigorously to the last.

A grayling butterfly pursues a black female model more determinedly than a model in natural colours. Not only this, but female models of a much greater size again produce many more responses than female models of usual size.

The adult European herring gull has an orange spot on an otherwise yellow bill, at which the chick instinctively pecks, causing the adult to regurgitate food. Experiment shows, however, that the chicks peck more vigorously at a *red* spot, and still more vigorously at a completely red bill.

The young of the Arctic tern instinctively solicit food from models shown to them of head-profiles having red bills and a strip of silver paper in the bill. This situation of course mimics the normal stimulus-picture of a parent returning to the nest with a fish. However, the experimental model found to elicit maximum response consisted of a *silver* bill containing a *red* fish – a complete reversal of the natural situation.

Finally, in an experimental situation, honey bees presented with a large variety of patterns are attracted maximally by the most complicated of the patterns. These patterns are more complex than those offered by existing flowers.

This selection of examples will suffice for the moment.

I can think of only two explanations for this phenomenon. The second of these I shall come to later. The first, now described, is inadequate, though plausible at first. It is that these responses (i.e. to the super-normal stimulus) are the residues of some earlier evolutionary situation. Perhaps, for example, oyster-catchers were once much larger and had larger eggs. At initial meeting, the notion seems not an unreasonable one. However, were this the case, and if the oyster-catcher has in time become smaller, then it did so because (and only because) survival favoured the smaller bird. The (hypothetical) larger birds, for whatever reason, had a higher casualty rate – so much so that they have now completely disappeared. The smaller birds that were left would, of course, tend in general to lay appropriately smaller eggs.

---

[4] He also adds: 'A closer study [of the phenomenon] might well be worthwhile.'

Now, matters in evolution are not altogether black and white. It is a question always of statistical tendencies, self-reinforced and gradually strengthened over time because 'losers' *tend* to die earlier, or to reproduce less adequately, or whatever. In the present case, however, certainly the eggs of a (small) parent bird inclined to respond optimally to large eggs would be survivally disfavoured over those of another bird responding optimally to small eggs – *the ones in general actually available* (since as indicated, and for obvious reasons, small birds tend to lay small eggs). Any 'large egg' response would therefore tend to be gradually eliminated, even if by any means it had somehow managed to survive the general down-scaling of the species. One might, indeed, look to find the *occasional, residual* large-egg response – just as occasional horses show very faint zebra stripes or occasional human babies have a small tail. But this is not at all what we have in the present case. We have a markedly aberrant response *throughout a whole species*. It is this fact in particular which both makes the behaviour 'remarkable', and which by itself effectively destroys the 'residual' explanation.

Let us at this point speculate rather wildly about the grayling butterfly mentioned above – and imagine that it somehow shot up the evolutionary scale overnight, and produced a cultural civilization. In that civilization – if at all resembling our own – would not the back-street pornographers sell to male butterflies stories of large, dark young female butterflies? Or, rather, would not the *front*-street bookshops sell stories of lady butterflies with a tendency to Mediterranean colouring, and a little larger than average, while in the back-streets the tales would instead be of monstrous, midnight-black creatures, and of male butterflies who made love to crows?

For do we not see in the human male's view of the 'ideal' human female one further example of an optimal response by an organism to a model other than that found in nature? I hope I have not jumped too quickly here. It will be recalled that in Chapter 2 we discussed the relative absence of body-hair in the human female, and noted this loss of the body-hair as a definite evolutionary trend. And we noted that in addition to hairlessness in the female, men also liked smoothness. We did not note there, but it is also true, that men admire women's breasts. Normal women seek to cater in a variety of ways for these various 'demands' by men, i.e. to improve on nature. And the fantasies of the pornographer, I suggest, merely take matters a stage further; so that we then find women dressed from head to toe in tight, gleaming leather, women with enormous breasts, and so on.

From the erotic illustrations of the pornographer, or simply the cartoonist, I think indeed we gain a very fair idea of the 'model' to which a majority of men would optimally respond. She has very abundant head-hair; large eyes with very long lashes; very full lips and a small chin; large breasts and large nipples on a relatively slender rib-cage; a very narrow waist, with broad hips and full thighs, small knees but full calves, with full, rounded buttocks; a generally fine bone structure (apart from the exceptions already mentioned), therefore fine arms, small hands and small feet. (One notes that clothes and accessories, where worn, accentuate these very features. One need hardly itemize false eye-lashes, waist-clinchers, the moulding brassiere, high-heeled shoes, and so on.) All the specific details I have listed are, I suggest, specific sign-releasers in the precise sense described by the ethologist.

Leaving aside for the moment an 'intention', if any, on the part of nature in these matters, the essential point cannot be denied: that, whatever else, the aberrant model (the over-large egg, the black female butterfly, the exaggerated human female, and so on) is *potentially* a powerful evolutionary force. For should it so happen that a black female butterfly ever appeared, its evolutionary triumph (through sexual selection) over the normal grayling female would seem assured. Indeed, as long as this aberrant tendency in the male butterfly persists, *must it not act as a selective mechanism tending to alter the colour of the female to black?*

The clincher to the proposal is that the proof of it actually exists in nature.

First, a case of circumstantial evidence.

According to my general argument, the young of the European herring gull, to take that example, should be acting as a selective mechanism on the bill-form of the adult – tending to cause it perhaps initially to develop a red spot, and perhaps ultimately a completely red bill. As it happens, the related black-headed gull of Eurasia and the laughing gull of North America do possess fully red bills. It is possible to argue, of course, that these other gulls show an older, not a newer, form of the general gull beak, and that the yellow/orange bill of the European herring gull is a later evolutionary product. In line with my earlier comments or residualism, however, this seems to me less likely than the reverse case.

Fortunately, what seem to me irrefutable instances of my still otherwise hypothetical position exist.

The young of many birds possess strikingly coloured patterns on the inside of their mouths, which are exposed when they gape for food. Experiment has shown that these patterns are releasers, which cause the parent to begin the feeding operation. As it further happens, these patterns are once again not of their optimal size (or intensity): that is, the parent can be shown to be more vigorously stimulated by larger (and more intense) versions of these patterns. The young *cuckoo* (who, as we know, is born and reared in other birds' nests) for his part possesses a relatively enormous mouth with large coloured flanges. This proves an irresistible attraction to the foster-parents, who then feed the young cuckoo to the exclusion of, or at least in preference to, their own offspring. The device is only one of several which the young cuckoo employs to defeat its foster brothers and sisters. What has happened is that the cuckoo (quite unwittingly, of course) has 'taken up the slack' in the parental response – thereby ousting the 'rightful' occupants of this particular ethological niche – who logically (and in terms of classical Darwinian theory) should be its most successful (i.e. *maximally adapted*) occupants.

There arises here the question as to what prevents the normal chicks from maximally occupying this niche; or in more general terms, what is it that prevents a releaser from attaining its optimal form?

We must, I think, envisage the various aspects and items of the individual organism's physiology/psychology as competing with each other, in much the way that the individuals of the same species compete each with the other. Every psychological attribute, so to speak, does the best for itself that it can, but its degree of success is limited (or at least may be limited) by other factors. Thus, say, the exaggerated plumage of a bird may be a powerful conditioner of the sexual behaviour of the partner, so that a *gross* degree of exaggeration results in many more than usual copulations and many more offspring. *But* there may well be a point at which the plumage in question begins to affect the quality or speed of the possessor's flight – or begins to attract a more than usual amount of a predator's attention. Thus, the attribute may also be (equally) *dis*favoured from another point of view. An ethological 'brake' will have been applied to the intended development under discussion.

In practice, I suggest, such 'brakes' are normally applied from a variety of angles. Where no such brakes happen to operate, however, that is, where no independent limits are set to a releaser's evolution, it will, over time, realize its maximum potential. As I shall suggest in the next chapter, possibly the enormous size of some dinosaurs resulted from just such a 'non-brake' situation.

From the general position I have been outlining, a somewhat shattering conclusion emerges. I think the case in evolution to be this : that nature first evolves a response – that is, first produces the inner situation – and then so to speak casts around for the stimulus to release it – that is, for the response, opportunity or environment that will fit it.

If this is true, the implications are formidable. For the traditional view of evolution is that of an *environment* acting on the organism. The environment (nature) is thought to call the tune, and the organism dances – if it can. If the organism *can* adapt, it survives.

Where it cannot, it perishes. So runs the traditional and currently accepted view of evolution.

The view I am in the process of expressing is wholly otherwise. What I am proposing is that the organism (somehow) of itself develops capacities and potential responses. I have for the moment not the slightest idea how it does this or why. Nor am I at all sure whether I judge the aberrant response to be born fully fledged, or whether it is somehow built up gradually. But this development seems *not* in response to environmental pressure or because of any kind of 'leading' in that sense.

I am, as I say, quite aware of the fundamental nature of what I am claiming, and of the many difficulties which my view raises. What it does do on the positive side, however, is to account for the otherwise inexplicable occurrence of the optimal response to stimulus-configurations which are not those found in nature. The 'drift' towards the 'ideal' form which this circumstance must continuously threaten to produce, seems indisputable. Whatever other views we hold in addition, we must add this force to existing ideas about the 'how' of evolution.

The notion of an already developed but stored potential is an extremely appealing one in connection with what has always seemed to me to be the weakest point of standard evolutionary theory – namely, that mutations and shifts in behaviour have apparently occurred just at the right moment on a very great many occasions in the course of evolutionary history – that is, *just* when the environment started 'acting up'. For myself, as for some others, this seems to have happened too often too conveniently. Now, if instead we have the notion of an organism with numbers of unused or stored responses – I will go into the question of how many responses shortly – there is a chance, I suggest a greater chance, of one of these providing an escape route than of the right mutation occurring to order at that precise point in time.

It should be mentioned that every response whatever has, of course, a range of variability. That is, a few individuals possess a particular response-readiness very strongly, a few very weakly, and the majority an average amount. This is true of both behavioural responses and physiological responses. A permanent drop in temperature will favour the survival of the few individuals who naturally possess a very thick coat (*and* in addition the lowered temperature will cause as a 'response' a thicker coat than the same individual would have grown if reared in a warm climate). Ultimately, only those animals with thick coats will be left. *But* – and this is the point I wish to emphasize – *the environment cannot produce reactions outside the organism's genetic range of variability*, outside what is after all 'there' in the organism. Beyond that some other solution has to be found – e.g. migration. *Or*, as I am now suggesting, the organism produces from storage some response it has never till that moment used, or outwardly shown any signs of possessing. It is to me an arresting thought that the aberrant responses of the animals we considered earlier might never have been acted out, and seen, in the physical, external universe (though nonetheless 'there' and in that sense existent), but for the action of the experimenter. It's a very telling argument in favour of the primacy of the inner life.

The question arises, how many stored or potential responses? Obviously one has at the present time little basis for any kind of accurate estimate – and this for more than one reason. For the sign-stimulus response-releasing situation is only one type or area of behaviour. Not all behaviour by any means is of this kind. We have, for instance, the whole field of purely learned behaviour. We have the field of exploratory behaviour, and so on. Some of the untapped potential that exists in these and yet other areas will be the subject of later chapters. That potential, however, is much less easy to demonstrate than are the 'aberrant' or additional behaviours of the so conveniently structured sign-stimulus situation – but this, I think, makes this demonstration all the more impressive when achieved. It is really a question of getting the organism to show what it can do. In general (and appropriately enough) it seems the situation must be one of life or death, and be *felt* by the organism to be one of life or death, for otherwise the animal, like ourselves, falls back on tried and trusted methods (i.e. behaviours). In man, necessity is said to be the mother of invention. Perhaps the same is true for all organisms.

However, to revert again specifically to the sign-stimulus situation, I believe that lower organisms, like human beings to a much greater extent, are 'haunted' by 'ideal forms'. The 'ideal form' is here defined as that one which produces an optimal response from the organism concerned. As we have seen, this is quite often not the form as it occurs in nature. While in general animals are very limited in respect of the creation of *artificial* ideal forms (though I think it is possible to say that, for example, the decorated nest of the bower bird and the collecting activities of jays and magpies have something of this behind them – points I shall take up later), they tend to take them when they can get them – for example when the experimenter provides them. Man, of course, makes his own ideal forms for himself. They appear in his fantasies and day-dreams, his drawings and stories.

I believe it is possible, as I have said previously, to distinguish between various kinds of archetype – for the archetype is, I propose, what we have been discussing throughout. The 'fantasy' archetype that is 'better' or larger than life is one that is still shaping or moulding our behaviour, or at least continually threatening to do so – towards, that is, the achievement or realization of the ideal form, the optimal releaser. The super-normal releaser of the ethologist is an instance of what I earlier termed the shaping archetype.[5]

On the other side, the archetype (or sign-releaser) that agrees with the model as found in nature has ceased to be able to move or mould our behaviour any *further* in any direction. It has got us, so to speak, where it wants us. For example, I believe the human baby is probably as baby-like as it can be – though only actual experiment could show that beyond doubt. I do not think there is a 'more baby' baby that we can visualize. There is very much, however, as I have argued, a 'more woman' woman – who could, potentially, exist; and who does, in a sense, exist in the sexual fantasies of males.

That we, mankind, can now artificially in a variety of ways achieve (or at least artificially produce) our own ideal archetypes – 'make our dreams come true' – without having to wait for actual, physical evolution to do the job for us, is an aspect of what I shall term 'symbolic evolution' – a notion to be developed later. Man can now, in respect of fulfilment, very largely 'do his thing' without the help of nature.

---

[5] A species of high evolutionary potential may be the species with a relatively large number of 'unnatural' optimal responses.

5
Darwinism: A Partial Equation

(i)

Darwinian evolutionary theory has the basic premise that stimuli impinge on the organism, that is, some aspect or aspects of the environment make a demand on the organism, to which demand it must make a response, in principle an adaptive response, or perish. The basis for the response-change is held essentially to be the occurence of a suitable genetic mutation. On this view, we see an organism fundamentally passive *vis-à-vis* its environment. The environment goes its unheeding way, and the organism so to speak stumbles along in its wake as best it may. Nonetheless, though passive, it is held that the organism is able to store or 'learn' genetically the lessons of its past vicissitudes. This circumstance then tends to result, over time, in both an increasingly complex organism and, moreover, one ideally suited to the environment in which it currently finds itself. As molten metal when poured into a mould perfectly assumes the shape predetermined by the mould, so is the organism a (fairly perfect) cast determined by the mould of nature and environment.

The two points on which Darwinian theory is weakest have long seemed to me to be (1) its reliance on suitable mutations occurring more or less just when required (that is, at the moment in time when the environment begins to undergo its sometimes fairly rapid changes) and (2) its view of the organism as possessing no kind of energy or drive (leave alone any trace of 'will'), but as being simply passive clay in the hands of its surroundings.

My own view, as already expressed in the previous chapter, is quite other. My proposal is that the organism *first* develops a response. This then, so to speak, either casts about for a suitable environment or is simply stored against the day of its possible use –

a day, incidentally, which may never come. Although this next point does not itself directly validate my proposal, it seems clear that an organism with six or ten or however many spare shots – that is, unused responses – in the locker is a more likely proposition for survival than one reliant on a favourable mutation occurring to order at a particular moment.

Linked with the (objective) situation I have proposed are, I further suggest, some extremely important subjective concomitants. I see at least some organisms, at least some of the time, and in some sense of these terms, driven or 'haunted' by an 'urge to become'. There is, I propose, some kind of inner urge or 'wish' to *realize* the stored potential. This proposed *subjective* concomitant (though not its *objective* correlate) is indeed not easy to demonstrate in animals – for of course we can have no direct access to an animal's mind, even if it has one. In one sense, then, any judgement of an animal's subjective state must always remain only a working assumption. Yet in man we do have ready access to subjective states. In man we have not only no lack of an overtly expressed and acted-out desire to become, but many verbal and fantasy clues and statements as to the inner state of the nation. To these we shall come shortly.

It is clear that under the impetus of any suggested drive or search to 'become', an organism (man or otherwise) would *encounter* the environment (limiting or external reality) very much as an opponent – which it either then defeats or is defeated by. Success in this fight, one must emphasize, can always at best be only relative (the ideal remaining almost inevitably somewhat out of reach – that being of course the connotation implicit in the term itself) while defeat may be absolute – as for example when the individual is killed, or a species becomes extinct.

My otherwise purely 'philosophical' views on the possible 'wish' of organisms to respond more vigorously than strictly necessary not only to actual situations but to situations as they do not even exist in reality gains very considerably in credibility, as we have seen, from the well attested findings of experimental ethology – though I would emphasize that these are not their sole support. As we noted, Tinbergen and many others have shown that in some cases the form of the optimal releaser of instinctive responses differs from the releaser as it is actually found in nature. I have chosen to call this 'unnatural' optimal releaser the 'ideal form'. In some cases the difference, or differences, between the naturally occurring and the ideal form is not great, and can indeed be measured to its actual limits. Thus the oyster-catcher optimally prefers a clutch of five eggs to the natural three. It shows less than optimal preference for still larger clutches. Yet the same bird, as it happens, prefers the very largest (single) egg with which it is provided, and always larger sizes have the effect of luring the bird away from its previous choice. Similarly, in terms this time of colour, the male grayling butterfly goes all the way possible to the extreme – the completely black female. This model is maximally preferred to the normal speckled female. In respect of size an apparently open-ended preference is manifested. The larger the female, the greater the preference.

It is I think beyond dispute that these hidden response tendencies (hidden in the sense that under normal conditions they are not revealed) constitute a potentially powerful evolutionary shaping force. Any chance deviation in the releaser in the direction of the 'unnatural' urges of the respondent will, as I have suggested, inevitably cause that deviation to be survivally favoured over normal forms.

Several points arise. How long may a deviant desire persist unused in a respondent? A thousand years? A million years? Or indefinitely, failing some genetic change in the responsive infrastructure? The last seems the most likely possibility. *Though never in fact once employed* (in itself an astonishing idea), the urge towards the 'unnatural' form could, I suggest, persist through entire geological ages, awaiting its possible moment.

A further consideration. Do we see in the 'unnatural' response some minor aberration on nature's part – in her otherwise well-ordered affairs – or are we seeing here the *normal form of all such instinctive responses prior to their shaping of the stimulus environment in the desired direction to the desired point*? May not the normal matching-up of optimal response with the normally occurring releaser *always* be the end-result (the as it were expended cartridge) of a response having realized or 'taken up the slack' in its potential?[1]

This view gains from the consideration of the human male's sexual response to the human female. As I have argued, there is an 'ideal' woman to whom the typical male responds more vigorously than he does to woman as we tend to find her in nature. The precise details of what the male 'ideally' likes are not far to seek, and are found most explicitly, as I have suggested, in pornographic

---

[1] I am inclined, as I said earlier, to wonder whether the enormous size of prehistoric reptiles was one striking instance of this process. While the humble present-day lizard or frog must content itself with puffing up portions of its anatomy, perhaps under conditions of superabundant food supply or relatively ineffectual predators, the early reptiles were able actually to execute their reptilian 'dream'. 'Largeness is all,' a reptile Browning might have said. It is further possible that, in addition to their vast actual bulk, extinct reptiles like some of their tiny descendants today also possessed inflatable sacs that could make them appear larger still.

literature. The same general tendencies are also reflected, however, in the normal woman's attempts to please or catch the attention of the normal male. I have indicated something of the direction of the wish-tendencies of the male and the corresponding response-tendencies of the female – for example towards longer legs (hence high-heeled shoes), more complete body-hairlessness and smoother skin (hence depilation and silk and nylon underwear), larger and more provocatively shaped breasts (hence the brassiere, both padded and shaping varieties) and so on.

Now, it can be argued that these aids to beauty are of very recent origin – and would this not suggest that the desires which produce them are of equally recent genesis? Not so. For as I have shown, we have good reason to suspect that such wish-desires can persist undiminished in strength over enormous periods of time, even though in no way revealed, let alone gratified. My guess is, then, that could we confront some lonely Cro-Magnonoid with Cyd Charisse or Marilyn Monroe, suitably attired for the occasion, that man would be as 'turned on' by the sight as we ourselves and 'blow his mind' along with many of his latter-day descendants.[2]

On a more serious note, legendary figures such as Lillith, Helen of Troy, Delilah, Salome and Cleopatra, and the deeds carried out in their names by men – together also with the central role which the beauty of women plays, and has always played, for the poet and artist – suggest how very, very significant is the ideal woman, how great the longing for her in men's hearts, in short how great a motivating force she has always been.

Aside from such precise 'ideals', we in any case find in human life a central role both assigned to and played by the concept of *general* idealism. This is the latent theme of virtually every public address and after-dinner speech, the inner content of all manner of creeds both secular and religious, and the very stuff of our tales of heroes and martyrs. In addition, it is commonplace to speak of the 'ideal father', the 'ideal husband', the 'ideal soldier', and so on – and all, at least intermittently, strive to be these. Is it possible to think of 'idealism' as the state of mind produced by an ideal image or, more probably, by the collective influence of large numbers of such images?

Many writers of stature have given very explicit expression to the notion of the abstract, over-all ideal – the state of idealism – wholly divorced from any specific (certainly any specific archetypal) ideal. Thus Robert Browning said, 'a man's reach should exceed his grasp, or what's a heaven for' and 'in heaven the perfect arc, on earth the

---

[2] Throughout this section I am really speaking of the A-dominant. Matters are more complicated in actuality by the admixing of B-dominance and by the B-dominant's sometimes differing inherited response tendencies.

broken round'. The artist Conti in Lessing's *Emilia Galotti,* dissatisfied as ever with his work, complains that in the mind the vision is perfect, but en route through arm and brush the perfection is lost. Goethe's novel, *Werther,* again is the story of a young man driven by all the feelings and aspirations of the artist, but lacking the technical skill to execute his concepts. His despair of not being able even to approximate to his artistic ideals leads him to suicide. In real life, the constant reproach which many artists bring to their own work (by normal standards far above what the remainder of us achieve), leading indeed sometimes to actual suicide, or to excesses of striving which are sometimes tantamount to this, mirrors in actuality the fictional situations I have just described. Nor are such degrees of idealism confined to artists.

I would define 'idealism' more precisely at this point as the attempt or wish to approximate to a model or models 'in the mind', whose attributes are other (are larger, coarser, finer or whatever) than those of the real-life situation or object.

This definition, as we see, is capable of embracing both human and animal behaviour – always granting, of course, that animals possess something approximating to mind or conscious experience as we know it (and always with the drawback that even if this is so we can never know for certain by any kind of *direct* observation what it is that the animal subjectively and consciously experiences, if anything).

Are there any apparent examples of idealized or idealizing behaviours in animals, apart from those artificially produced by the experimenter-ethologist? I believe there are. I think that Jane Goodall's chimpanzee, Mike, was showing himself as a more ideal leader (both in his own estimation of what constituted leadership and himself as a leader, and in the eyes of the other chimpanzees) when he banged paraffin cans together to enhance his charging competitions with Goliath. As Jane Goodall remarks: 'It seemed that Mike actually *planned* his charging displays – almost, one might say, in cold blood. Often when he got up to fetch his cans, he showed no visible signs of frustration or excitement – that came afterwards when, armed with his display props, he began to rock from side to side, raise his hair, and hoot.'[3] And when the courting bird puffs himself up to twice his size is he not, as I have already suggested, trying to be the bird the female would have him be – the much larger bird? There is no lack of examples of *this* kind of 'idealism' among animals. However, the first example just now was definably different, in that it involved the utilization and incorporation of objects from the material world – such as man himself

---

[3] Jane van Lawick-Goodall, *In the Shadow of Man* (Collins, London, 1971), pp. 110-15.

practises when he puts on an ornate helmet or a resplendent uniform.[4] I do not doubt that enough examples of such practices will be found among both primates and other animals, once we know what it is we are looking out for. These instances may well of course not always be very impressive in our eyes.

One further instance seems certainly offered by the bower birds. The males of this group of species modify their surroundings in order to influence the female. Usually the males themselves have relatively drab plumage, and in general the more dull the plumage, the more elaborate and eye-catching the bower they build – a reverse ratio of fairly evident significance of course. What the male bird precisely does is to build a tent-like structure over the nest and surround it with a 'garden' (sometimes an avenue) decorated with bright berries and flower petals, including jewellery if he can find it.

The specialist literature does not doubt that the functions of the normal bird's bright or exaggerated plumage have here been externalized and applied to the nest. That is, what the average bird achieves with its plumage, the bower bird achieves with his architecture. The aspect I would like to emphasize is that while in the case of normal birds it can readily be argued that they have no cognizance of their own plumage, much less any hand in its creation, it is far harder to argue that the bower bird is not in some sense subjectively aware of the objects he handles in building his bower. Moreover, as far as I am informed, each bird makes an individual pattern and a 'personal' choice of objects – a further most interesting development away from the normally much more rigidly stereotyped, instinctive action.

In all the varied examples I have discussed, both human and animal, it seems to me we see an organism very much involved in an attempt to become. An organism not content simply to *be* (as Darwin would essentially have it) but one, in a sense which does no essential violence to the meaning of these terms, striving to 'fulfil' its 'potential'. In this general context I would certainly adduce the very impressive evidence supplied by Jesperson which shows the personality of the pre-speech child striving (and, indeed, striving joyously and positively) to break through to, and to master, its social and material environment. One cannot hold back, also, from considering here the achievements of severely handicapped individuals such as Helen Keller and Christy Brown.[5]

That the various forms of the 'push from within' (not all necessarily archetypal) are at least partly underpinned by hormonal,

---

[4] Chimps, of course, *regularly* use chunks of foliage and branches to enhance their charging displays.

[5] *Down All The Days* (Secker & Warburg, London, 1970).

maturational and other such physiological factors I do not dispute. I am concerned, however, with the *subjective* concomitants of this (possibly) biochemical infrastruture. I am concerned most especially with that *psychological* entity, the personality, which appears at least at times, and at least in some areas, to be able to act in its own right on its own behalf.

My position is at variance with that of the classical evolutionist not in the sense of challenging the details of his observations as such, or the role he ascribes to the environment, but in my assigning to the organism both a positive role and a measure of self-determination.

By way of analogy, in the use of cutting and grinding mechanisms of all kinds some method is employed of holding the object to be treated firmly against the cutting or grinding edge. Without this, the operation is very severely reduced in efficiency, if not entirely abortive. In the typical household meat-grinder, for example (as well as in many commercial operations), the force employed is simple gravity.

This notion leads us to another analogy, which is perhaps closer to a model. A geologist might be concerned to study the flow of rivers, their effect on the terrain over which they flow and in turn its effects on them. In so doing he need not concern himself, either theoretically or practically, with the concept of gravity. Rivers *do* flow downhill – and with this much given, the geologist can go about his work. Yet, of course, without the *fact* as well as the effect of gravity, the river would not flow in the way it does. I believe, similarly, that without what I term the 'energy of evolution' the evolutionary processes that Darwin studied would not be there to study.

The 'energy of evolution', analogous or parallel to the notion of gravity in the physical sciences, is to be understood to be that force which holds the organism against the cutting edge of the environment.

(ii)

My general contention, then, in psychological terms, is that *response* – that is, the ability to make the response – precedes not necessarily the existence of the stimulus as such (though that sometimes), but certainly its effect on the organism. Conceptually and literally, the response exists prior to and independently of any environment or any stimulus situation in which it may subsequently be manifested or acted out.[6] I base this general contention in the first instance on

---

[6] Of course, I have stated this very simplistically. First, I am not suggesting that the response itself has no evolutionary history, or that this

the evidence supplied by ethologists in the context of sign-releasers and the fixed, inborn responses which these elicit – for this evidence seems incontestable – but not solely on that evidence.

There follow now examples of two actions by animals which appear to go beyond the limits of what are normally considered their capabilities. As in the sign-stimulus situations already examined, the animal here produces an action (but this time not merely a *degree* of action) never, as far as we know, hitherto performed by it: an action of which it must, naturally, in some sense be and presumably for some time have been capable; but an action which we as observers could not have predicted in advance nor, as already said, of which we would have even suspected the animal to be capable.

The first example is taken from Konrad Lorenz's book *King Solomon's Ring*.[7]

It appears that Lorenz at one time had a pet raven, which he called 'Roah' – this word being an approximation of the call-note of a young raven. This raven, remaining even as a fully grown adult a close companion of Lorenz, was always distressed to see him walk in places which the raven considered dangerous. Ravens as a species have a method, instinctive in origin, for issuing to another raven on the ground an invitation to fly. The method consists of flying in from behind, close above the other bird, at the same time wobbling the closely folded tail and emitting a series of sharp sounds – which Lorenz renders by *krackrackrackrack*. On the occasion described, Lorenz was walking in a 'dangerous' field. The pet raven flew over his head from behind in the manner just described, at the same time looking back to see if Lorenz was following, and emitted not the call-note appropriate to the occasion *but the word 'Roah' with a human intonation*. The inference is clear. Far from realizing that 'Roah' was his own, that is, the raven's own name, Roah had 'understood' this to be Lorenz's call-note. The raven had used Lorenz's 'name' (i.e. call-note) in an effort to get his attention.

This is, as far as I know, the only recorded instance of the use of

---

evolution may not have been affected by environmental (i.e. stimulus) factors. And of course, if the response is to be actually *used* and if we are to *observe* the use of the response, some kind of environment is essential as part of the full equation. These aspects are not in dispute. What we are discussing here is primacy. Again, in the question of primacy I seek not at all to deny that the material world (in all probability) in some sense exists in our absence, and therefore also *prior* to our existence. However, an aspect of that physical, external world becomes what we call a stimulus *only* once an organism exists to receive and to react to it.

[7] Methuen, London, 1961.

spontaneous, meaningful language by an animal. But perhaps still more importantly, the bird had broken up one of the chains of inherited instinctive behaviour, and substituted a piece of behaviour of his own devising at the end. The implications of this are, I think, shattering for any psychologist.

I personally see no way of explaining the bird's behaviour, other than by some process of understanding on the bird's part, such as suggested.

The second example is taken from the London edition of the *Evening News,* May 18th, 1972, and quoted verbatim.

> RSPCA officials today [Thursday] took off their hats to Tim, a Jack Russell terrier. For the dog saved its own life after rescuers had been battling since Sunday to free it from a crevice in a mountainside. Tim was trapped 60 feet down ... Then the RSPCA inspector tried a new tack. He borrowed the cap belonging to Tim's owner, Mr Danny Evans, and lowered it into the crevice tied to the end of a rope. 'Amazingly Tim had the good sense to bite on the cap and he was gently hauled out,' said Mr Mills.
>
> The battered green, tassled cap that helped to save Tim was knitted by a friend's wife four years ago. 'Danny is never without it', Mrs Sylvia Evans said.

I am unable to account for the dog's behaviour in mechanistic terms. It is perhaps possible to imagine that Tim bit the cap as a reflex or hysterical action. But that he held on when lifted clear of the ground for 60 feet cannot, I think, be so understood. To hold on, no doubt painfully, throughout a manoeuvre lasting some time ('was gently hauled out') requires, I suggest, that the dog well appreciated what was happening, and what it was necessary for it to do. Of course, the fact that it was his owner's cap no doubt helped enormously – general conditioning processes certainly aided rather than hindered the act. No doubt, too, as a pup Tim had swung free of the ground in some playful tug-o'-war with his master. Yet, it seems to me that for him to hold fast during the lift up over a considerable distance is a completely new behaviour, one that came fully fledged into being at that moment, and was, of necessity, put into use without prior testing – unless then conceptually, in the animal's mind.

Both the cases just described involved a 'life or death' situation – a crisis calling for some kind of transcending effort. This, it seems, was the stimulus which caused the two organisms to act to their limits – in a sense, beyond their limits.

Roah, Tim the dog and Mike the chimpanzee, all three show us behaviours of which we did not suspect these animals to be capable, before the manifestation of those behaviours. These, on this occasion, are not already fixed responses, existing beforehand as precise genetic encodements, but are new responses made up, and indeed tailor-made-up, on the spot, by the organism, to meet the unique requirements of a unique problem situation.

In respect of responses, organisms, it seems, are capable of two important things. First, of 'storing up' or carrying of unused, inherited responses – which, as far as we can tell, have never been used in the whole history of the species concerned – against the day when they *may* be needed, or at least become possible. (These responses await so to speak the other half of the complete pattern – the stimulus that will release them. May this, perhaps, be one of the meanings of the symbolically expressed 'magic spell' or 'open sesame'?) Second, they are capable of inventing *entirely new responses*. In this second case also obviously the existence of some kind of general response-*potential* must predate and underpin the response – for of course elephants cannot swim under water and crocodiles cannot fly – so that here again questions of phylo-genetic history and background are involved. Nonetheless, the response *as such* is new, is a *variable commodity,* and despite *general* phylogenetic determination is not itself wholly predetermined. As in a game of chess, the rules determine only partly the nature of the next move, and the move itself is unique. Or, as in musical composition, the composer takes notes many times used and from them makes a new melody.

Under both headings – the strictly genetically encoded, but never yet used response, and the newly invented response – an organism may, and indeed does, make a response of which it (that is, its species) has been in some sense capable for perhaps hundreds of thousands of years, but has never demonstrated till this day.

This point enables me to deal with a criticism of my general position made by the late Sir Cyril Burt. He asked, rather angrily, whether I thought the brain of modern man differed physically from that of man 5,000 years ago, and, if it did not, how I could justify the use of the term evolution, in respect of changes in man's behaviour that have occurred during that time. My belief, actually, is that the brain of man 5,000 years ago probably did not differ from the brain we ourselves possess. I would go even so far as to say that the brain of man 20,000 years ago did not differ. What is, however, clear is that man was then and is still today *learning the use of this brain* – this organ which actually came into *physical* existence, in its present form, perhaps a very long while ago.

The notion that a structure of this complexity (or more precisely, the upper layers of this total structure) could have come into existence before a complete use for it existed is I think a wholly daunting one. As I see it, it seems that we (that is, consciousness) are continually 'moving into' areas of our brain as it were prepared against our coming; and when we put these parts of the brain to use, they function, it seems, faultlessly. It is as if a person with no knowledge or experience of metal, let alone of metal under stress, of aerodynamics or electronics, were to have drawn up detailed plans and specifications for the Concorde airliner to be built 20,000 years later – and so well that when the necessary materials and facilities were available for building, the plans would be found to be entirely accurate. This is of course by no means a perfect analogy. For in the case of an organism it can be assumed that any pre-prepared physiological and psychological structure would somehow tend to encourage the organism along the path which would necessarily end in the structure. That point, however, does not really make the (alleged) prior existence of the structure any less astonishing.

It is not my intention to make these remarks sound in any sense 'religious' – I am merely describing what seem to me to be the readily observable facts of an actual situation. I would, however, suggest that certain aspects of the conventional religious view possibly derive from an archestructural perception of the situation I describe.

When man first began reading, he did so by speaking the words out loud. In time this became muttering, with slight lip movements. (Substantially this same process is gone through by modern youngsters in the process of learning to read.) One or two individuals in the past, however, at some point realized or otherwise discovered that reading was possible without any sound at all (without even sub-vocal sound, such as, with special equipment, we can detect in the poor reader today). *True* reading is entirely independent of the spoken word – the eye so to speak 'gulps' chunks of text whole, and the sense is transmitted direct to destination. This, then, was an ability which man discovered in himself (discovered that he was able to do) but which was always there – 'always' meaning since mutational inception. How much more impressive again is the discovery in ourselves today of the ability to perform mathematical calculations, or to compose music. Yet, at least individual members of Neanderthal and Cro-Magnon, I would suggest, had – *must* have had – all these abilities, that is, all these response *potentials*. As things stood, however, they had no, or almost no, chance to demonstrate them.

In sum and in particular, I suggest that mechanistic accounts of evolution – such as propounded by Darwin, and so very useful then, as now – when applied without at least equal regard to the inner functioning and nature of the organism, are no longer a help, but constitute a very serious block to our understanding of the real nature of evolution.

6
Symbolic Evolution

A question which the reader might like to ponder at this stage is whether ideas are organisms or objects; and if neither, to which third class they might be assigned. My own view is that ideas do not belong in the last analysis to either of these classes, and we shall be concerned with that general proposition later. For the present we shall be discussing what can be understood about ideas by considering them on the one hand as organisms, and as objects on the other.

My view of the idea as organism was, initially, not more than merely analogous. That is to say, while reading evolutionary texts I was several times impressed by apparent parallels between the behaviour and nature of organisms and the behaviour and nature of language, and further, the nature and behaviour of concepts. The animal behaviours in question would, it occurred to me, make useful analogies in illustrating certain of my feelings on linguistic and other mental phenomena. But as further time passed I came to feel that the situations I was considering were more homologous than analogous; and in the course of what follows I shall hope to reproduce at least some of that conversion in the reader.[1]

From a purely general, let alone from any specific, point of view to treat ideas as organisms rather than objects seems the inherently more reasonable position. For, after all, ideas are produced by organisms and are found only in association with organisms. Only

---

[1] The concept of *language* as a natural, organic phenomenon is by no means new. In the nineteenth century, philologists (e.g. August Schleicher, Franz Bopp, Rasmus Rask) took this view virtually as a matter of course. However, I personally find their discussion of the issues often not better than merely figurative. Nevertheless, their intention is of great interest. They were, in my terms, attempting something like a Self-description of language.

organisms can use and understand them. Indeed, without organisms they do not exist at all. A computer, as we have seen, does not contain, produce or apprehend ideas. (It produces something, of course – that is, marks on pieces of paper, or whatever – which we, and only we, then recognize as ideas.) So a building, once every human being on earth is dead, is not a building. It is only a clump of stone and metal. Or rather, it is only a clump of 'stone' and 'metal'. Or rather, 'a' 'clump' 'of' 'stone' 'and' 'metal'. I trust the point is clear here: there is no word we can utter that is not anthropomorphic. Similarly again, the policy of the Northern Europe Presbyterian Building Society (which is itself a mere idea, since no such society actually exists) has no existence when all the employees and clients of that Society are asleep.[2] To this last it could be objected that the policy then *does* still exist, because stored as an (at that moment admittedly inert) memory trace in the brains of the sleepers. In this I would concur. But I would ask whether there was not merely a great, but a quite, quite fundamental, difference between an inactivated and an activated memory trace – that is to say, a trace made conscious.[3]

What, in any case, to turn from generalities to particulars, are some of the alleged parallels between the behaviours of ideas and the behaviours of organisms?

Chapter 3 of Darwin's *Origin of Species* discusses the struggle for existence, and in that chapter Darwin makes the general point that this struggle is at its most intense between individuals of the same variety, and then less between varieties of the same species. Thus mistletoe can be said only in the most derived or metaphorical sense to struggle or compete with the apple tree on which it grows and feeds – for the death of the apple tree is no part of what the mistletoe seeks, even if it sometimes inadvertently causes this. In a less metaphorical sense – a more literal sense – mistletoe *can* be said to compete with other fruit-bearing plants, say, in its attempts to attract birds, for it is by the agency of birds that such plants distribute their seeds. Other species and varieties of mistletoe would here

---

[2] It will be understood that I am putting only one side of the case for the moment. It is, of course, from many points of view useful as well as meaningful to treat the policy of the Northern Europe Presbyterian Building Society as if it had some independent, objective existence. This is the (in that context acceptable) process of reification – of regarding an abstract idea as an object.

[3] I would, further, ask what activation is – and whether in the process of activation, and in the description of that process, we have the only fact of real significance not simply in our own existence, but as far as we are concerned in the existence of the universe.

be the most direct competitors because of their great similarities with the individual we are considering. But mainly, and in the most direct of senses, mistletoe competes with other individuals *of its own variety*. For at any growth site there can be at best place and nourishment for just so many, and in the most extreme cases only for one.

Avoiding definition of terms for the moment, is it not equally true that in the realm of ideas 'individuals' and 'varieties' of the same 'species' compete most vigorously with each other – where the struggle for survival is most intense? I am aware that I am here begging my own terms of reference. But let me persist for the moment. For if ideas are organisms, then it would follow that idea-systems are species (or genera, or whatever).[4] Let us assume that religion, science, philosophy, politics, and so forth, are species; and that sub-divisions within these are varieties. On the basis of this tentative assumption, and by way of verification of it, we look for more conflict between the varieties than between the species. This indeed is what we find. For during the Thirty Years War in Europe, or at the present time in Northern Ireland, we find actual outright war occurring between Catholics and Protestants (that is, between the *varieties* of religion). Or in Spain in the 'thirties, and in Vietnam today, we see actual outright war between *political* varieties of ideas.

On the other hand, while it is true that no love is lost between priests and scientists; between priests and politicians; between politicians and scientists and so on – and while it is true that occasionally an individual member of one of these groups is killed by members of another – wholesale, outright war we nevertheless do not find. The 'species' here clash less dramatically than the 'varieties'.

The analogy stands up in yet further detail. As Darwin points out, though there is less competition between species than between varieties, there is nevertheless relatively more competition between more closely related species than between less closely related species. Politics and religion, I would argue, are more closely related to each other than are politics and science, in terms of the kinds of phenomena they consider and handle, the social areas in which they operate, and so on. We indeed find more direct, overt clashes between the first two species mentioned than the second. One has in view the confrontations between, say, Catholics and Communists in Italy, France and elsewhere, producing occasionally rioting and actual deaths. The (in any case rarer) clash of politician and scientist does

---

[4] Perhaps the subjects of the school and university curriculum. It is worth asking from whence we derive these – for in nature many, if not all, of the phenomena concerned (which we separate into subjects) are simple continua. May it be that we archestructurally or unconsciously group these into species?

not produce overt manifestations even of this reduced magnitude.

I suggest that a further index of the distance between the differing 'territories' occupied by any two (or more) idea-system species is the ease with which any individual can be a simultaneous, functioning member of each. Thus it is fairly easy to be both a politician and a scientist. It is less easy to be a politician and a priest – that is to say, there is in this second case a much greater conflict of loyalties, more frequent cause for heart-searching, more need to *define* the boundaries (the spheres of influence or territories) of the two idea-systems. Such territorial crises do not tend to arise either so often, or so intensely, for the one individual-as-scientist-and-individual-as-political-activist, though they *can* and occasionally do, for example in Soviet Russia. These conflicts perhaps rather more readily arise for the one individual-as-scientist-and-individual-as-congregant, though that they need not necessarily is shown by the harmonious relations which at least at one time prevailed between church and science; and by the fact that eminent scientists are, occasionally, also deeply religious.

So much for 'varieties' and 'species' in ideas. What of the individual, single idea? We might here consider as an example the wave and particle theories of light in theoretical physics. Each has something to recommend it – enough, at least, to ensure its survival for the moment. These two ideas as it were confront each other directly, with no intervening distance. They occupy, potentially, precisely the same space. They are like two evenly matched plants growing next to each other in a space that is ideally meant for one. A slight improvement in the explanatory power of either would result in the weakening and death of the other : just as in the case of the two plants, any increase in the amount of shade, or the alighting of a parasite, on one of them would give the other that little extra leverage – so that at first slowly, but always more surely, and finally rapidly, the other would die.

The example we have taken concerns, as it were, two 'adult' individuals – two evolved ideas; and in the parallel case two, say, oak trees. What of the *seedling*, the individual idea at the moment of conception or birth?

The seedling situation we find, for example, in any classroom or seminar debate. Assuming the people involved are not emotionally aroused or otherwise neurotically incapacitated, ideas are produced, discussed, accepted (temporarily or permanently) or defeated. An idea might survive for fifty seconds ('I think you'll find that Smith has already faulted that view') or ten minutes, or several days, before some fatal flaw is discovered. It may survive sufficiently long or well to be inserted in the permanent records of the study-group. And whether it survives fifty minutes or fifty years it may in that time have succeeded in giving birth to other ideas. The longer the survival and the more 'offspring', the greater the chance of an idea becoming the original ancestor of a 'strain', 'variety' or even a 'species' of ideas.[5]

A further general parallel is as follows. *Actual* fighting in animals occurs least between individual members of the same variety. This is extremely beneficial to the variety from an evolutionary standpoint – and presumably therefore why the situation is so, for a damaged organism would naturally stand less chance of survival. Instead, actual fighting has been replaced by the largely instinctual, complex, ritualized movements and signals (the baring of the teeth, the bristling of the fur or feathers, and so on) of a purely notional battle. In a short time the symbolic 'battle' is decided – one animal emerges as victor and the vanquished individual withdraws. Only exceptionally are actual, and still more rarely lethal, blows exchanged.

The signals in question are in large measure variety-specific. While closely related varieties still normally possess a sufficiently related repertoire of signals to make actual fighting unlikely, nevertheless as the gap between varieties increases more *actual* fighting will tend to occur, because the animals lack the common symbolic means of preventing it.[6] At the same time a counter-tendency is at work: for as varieties grow more dissimilar, so the 'areas' they operate in (the type of food they eat, the prey hunted, the sites selected for mating and breeding and so on) also begin to differ. Increasingly less *chance* of conflict is therefore now likely because the interests of the animals are increasingly divergent. This is one of the reasons for the reduced

---

[5] At the outset of this paragraph I said that I assumed emotional maturity and the absence of neurosis among the people concerned. Very often such is not the case, and we have the situation that ideas which should survive, because they are better ideas than existing ones, are shouted down and destroyed. Occasionally ideas are lost for ever for this reason. Surprisingly often. however, despite all the hazards of this kind, ideas nonetheless do survive in the long term on the basis of their merits – of their genuine fitness.

It is clear, I trust, that I am not unaware of the complexities of the kinds of situations I am using as analogies and moving through so cheerfully and rapidly. To open up a discussion of the reasons for and the mechanisms by which ideas survive, or to study the precise nature of the 'jungle' into which they are born, is not part of my purpose in this chapter. One or two comments on that subject will however be made later in passing.

[6] Compare 'fighting' between chimpanzees, and earnest fighting between chimps and baboons.

conflict between species, the still further reduced conflict between genera, and so on.

Thus, to summarize, there is this gradient: little actual fighting between the members of one variety; some actual fighting between divergent varieties and closely related species (where face-to-face confrontation occurs without enough of the controlling influence of inherited ritualized behaviour); and little fighting of any kind between widely divergent species, genera, and so on.

A parallel situation pertains from several points of view in the academic and scientific world. 'Competition' between individuals in any one branch of study is, for example, heavily ritualized. (There is, of course, no suggestion whatsoever that *these* are inherited behaviour patterns.) The principal methods of publicizing one's ideas (of 'fighting') are by open debate at discussion seminars, by the publication of papers in learned journals (the form, layout and length of these papers being usually also firmly prescribed), the reading of papers at conferences and seminars, and the submission of theses. On all these various levels personal abuse is heavily frowned upon – indeed, a deliberate aim in these matters is the avoidance of all emotion and personal involvement, in favour of objectivity and reasoned argument. Losing one's temper is already a serious failing; the actual coming to blows unthinkable. How like, then, in their *style* and *effects*, are the protocol of the academic, and the rituals of the animal, world.

Where in the academic world *do* we get loss of temper, personal abuse and the breaking-off of relationships? Why, precisely (a) at those points where common lines of inquiry begin to diverge into 'sub-varieties' and potentially full 'varieties' and (b) again where the same subject matter (as it were the same nest-site or feeding ground) is invaded by a more distantly related discipline.

Instances of the first case are the well-known estrangement of Freud and Jung; and the divergence of the Russian and American conditioning theorists (i.e. Pavlov *v.* Skinner). Of the second, the clash among psychiatrists, psychoanalysts and learning theorists on the nature and treatment of mental illness; and of genetic and environmentalist psychologists on the subject of intelligence. *Still* further divergence, however, rapidly reduces conflict. In general biochemists do not quarrel with learning theorists (even though the former may be concerned with the chemical basis of learning) or mathematicians with biologists. The distance between them is adequate.

Turning now to language, since Darwin a well-accepted feature of organisms is the circumstance that the embryonic, the young and even the adult form of the individual show traces of organs and functions long since superseded and replaced. These 'rudimentary structures', as Darwin termed them, provide important clues to the evolutionary path which the ancestors of the present organism once traced. Three instances of the phenomenon from the myriad number available are : the well-known gills which the human embryo briefly possesses during its early life, the inherited upper teeth of the young calf which, however, never cut through the gum (and which indicate that some ancestor of the calf once possessed such teeth) and the shrivelled stumps of wings found under the now solid carapace of some beetles.

An extremely important corollary to this basal situation, one that we frequently overlook, is that the rudimentary structures tell us that once the ancestor possessed that organ or function *in a fully developed form*. If once we accept that language and idea are homologous, and not merely analogous, to organisms, then we are entitled to deduce ramified and at one time dominant social and behavioural structures from, for example, very slight linguistic clues. An example follows immediately, which bears out the assumption here.

The modern English word 'left', derives from Anglo-Saxon *lyft*, which meant weak, worthless, womanish. We have, as it happens, consciously forgotten that meaning.[7] We have also forgotten that 'cack-handed', a word applied to left-handedness especially in the North Country, means literally 'shit-handed' (Latin *cacare*, to void excrement, and general Indo-European *kakka*). Nonetheless, to the etymologist (as to the biologist and embryologist in their fields) the traces of the former attitude of society to left-handedness here are clear. What is more, these were obviously universal attitudes, since otherwise the 'public opinion' of the time would not have tacitly (or rather vocally) approved the use, by in *fact* using the terms. The same former (and sometimes still residually present) attitude to left-handedness can be seen in the terms for 'left' in all the Indo-European languages (at least, those with which I am personally familiar), so that there are grounds for believing that that attitude to leftness was both general and deep-rooted among the Indo-Europeans as a whole – whoever and whatever they may have been. This was, I suggest, in origin an archetypal response – and more, an at that time still living archetype. In this *particular* example – if the case is as I suggest, quite aside from the *general* point that words, like organisms, bear the traces of their evolutionary history – we see that there is a much more directly biological, because archetypal, basis and content to at least some words than is usually suspected.

Though we must be careful not to lose the general thread of this

---

[7] That is not to say that the nervous system has necessarily 'forgotten' the original situation – as we shall observe.

chapter by entering too deeply into side issues (they are nonetheless important), perhaps some comment is in order as to precisely how a particular word is chosen ('survives') and 'evolves' when a new usage is required, that is, when some new situation arises which needs to be described and referred to. I do not suggest that there are never conscious elements in such a choice – and here in any case we must distinguish between the choice made by some influential idividual and the 'choice' made, that is, somehow sanctioned by common usage. Even when the choice is apparently pretty conscious, I still, however, suggest that unconscious motivation is very often, perhaps invariably, at work below – for instance, in the choice of the word 'left' to describe the Socialist movement. In the case of a popular decision – always by far the main influence in the evolution of language, in recent times somewhat augmented by the deliberate efforts of the writer and the academic – I believe the determinants to have been almost entirely unconscious or archestructural. One example follows.

We have in English three verbs virtually identical in meaning: 'level', 'raze' and 'flatten'. Depending to an extent on the precise context, my own impression is that the word 'raze' is the one most commonly used. According to the *Shorter Oxford Dictionary*, raze/rase was the first of these three words to be used in the sense of destroying or removing a building. All three words existed in other senses at the time, and the other two roots are absolutely older. The choice of raze therefore was in no way forced, except in an unconscious sense. This I see on the one hand as part of a general tendency to avoid words with Anglo-Saxon roots in public language, in favour of Latin and French derivations. More importantly I believe that the choice here is and was determined by the identical sound of a verb of opposite meaning, that is 'raise'. For it is the case that as we go back in time the employment of one and the same word for a quality and its opposite increases in early languages: and, perhaps more significantly, young children repeat this behaviour at an early stage of their own language-learning.[8] Traces of this practice, and what look like occasional new examples, are sometimes found in the English of today.[9]

The scientific and currently fashionable approach to language is to treat words, and indeed all psychological phenomena, as objects – that is, to study them by the impersonal, objective methods applied so successfully to the world of physical objects. Information theory is one of such attempts.

---

[8] See, for instance, Otto Jesperson, op. cit., p. 120.
[9] See *Total Man*, Chapter 3.

I do not wish to spend too much time considering this side of the total coin, for it receives already more than adequate attention in our one-sided culture. However, if, say, the written names of the days of the week are examined, it is found that a good deal of each name is actually redundant in terms of the information conveyed. Already the abbreviations M., Tu., W., Th., F., Sa., Su., provide all the information we require to make no mistake about which day is intended. Still further redundancy exists, however. For the capital letter M is unmistakably recognized already at this point N. No other letter can be formed once we have proceeded this far. The capital letter W is unequivocally rendered by W, and capital S need not be continued beyond the top curve, thus ⌒.

Clearly, this kind of material constitutes a topic of interest, one of not only theoretical but very practical value — for example in the field of communications, where time, cost and accuracy are important.

I repeat, then, that I have no fault to find with objective approaches to language — which, of course, include variously also grammatical and syntactical studies, and so forth — *as such*. My objection is when these methods are considered an adequate main approach to language, let alone an adequate sole approach. There is so much in language — and, of course, that of which language is an expression, the inward, experiential life — that can never be grasped or accounted for in those terms. Some part of this is well expressed by Elizabeth Barrett Browning in the lines which follow.

> Women know
> The way to rear up children, (to be just)
> They know a simple, merry, tender knack
> Of stringing pretty words that make no sense,
> And kissing full sense into empty words,
> Which things are corals to cut life upon,
> Although such trifles : children learn by such
> Love's holy earnest in a pretty play
> And get not over-early solemnised . . .
> Such good do mothers. Fathers love as well
> — Mine did I know — but still with heavier brains
> And wills more consciously responsible,
> And not so wisely, since less foolishly.
>   ('Aurora Leigh', 10)

I would be willing to base my whole case against the objective study of language as a total approach on these lines alone, and personally consider the case won. Indeed, every time one reads these lines new insights and unfoldings into the nature of language appear, and such indeed is the unmistakeable sign of the artistic process.

What Elizabeth Browning is describing here is a view of language as seen or experienced by the Self, while in the later part she gently criticizes something of the Ego's approach and attitudes in this area. (For, once again, as in the whole of the present book, we here see reflected, and are discussing, the enduring *duality* of the human organism.)

Certainly I consider that neither the Ego nor the Self alone can give us a wholly adequate account of any human phenomenon – since all human phenomena, as we saw in Chapters 1 and 2, and however residually in some cases, contain elements of both. In making this comment, I should perhaps emphasize that I am not therewith upgrading the mass of human behaviours to System C or 'Person' status. That is a rare and special synthesis of the polar constituents, such as we find in the great works of art. Most human activity is compised of, and emphasizes, mainly one aspect of the basal personality – be it Ego or Self – and seeks frequently, in the last analysis always in vain, to eradicate or maximally play down all trace of the opposing *persona*. In the case of words (and perhaps of ideas in general) I believe, however, that the Self can tell us rather more. The reason for this is that, as I see it, language originated with the Self.

My grounds for this stand are both numerous and varied, and I do not intend to open out all of them here for fear of leading the discussion too far into this special issue. From experimental psychology comes, for example, the finding that girls on average begin to speak before boys, show fewer speech defects (all kinds) than boys, and learn foreign languages more readily both during childhood and in later life. In folklore, too, women are habitually characterized and stigmatized as gossips and chatterers, unable to resist the temptations of verbalization.[10]

Yet I would like to turn to 'evidence' of a quite other kind. As is well known, the Gospel of St John begins with the statement: 'In the beginning was the word.' This is perhaps one of the most discussed (and publicized) utterances in Christendom – as may be recalled, it was this text which Faust began to translate in his study – having almost the force of another Genesis. I believe the interest which this utterance has engendered is not fortuitous or of small significance.

---

[10] Connected, I think, with the greater appeal which relationships and social intercourse appear to have for women. This, as I hold natural, tendency encourages their more rapid acquisition of the means of communication.

My general view of such books as the Bible (the Koran, the Veda, the Upanishads, the Torah, and so on) is that they are fragmented, distorted and bowdlerized versions of older and perhaps once complete accounts – and each survivor may of course be a compendium of the fragments of many such – of the universe and the nature of life *as perceived by the Self*, often more or less heavily overlaid with the conscious and intellectual ethic of later periods. The more recent the text, the more conscious elements it tends to contain.

In support of my general claim here we have, most fortunately, one text from China which appears to have survived relatively unharmed (and though added to in historic times, not significantly distorted) from some very distant, I suggest prehistoric, past – the *I Ching* or *Book of Changes*. This is, literally, a miraculous book. Both it itself, and the treatment of acupuncture which is related to it, are today gradually becoming recognized as valid, but possibly non-causally based, forms of knowledge or experience.

However, we are presently concerned with St John. Staying for the moment with the literal meaning of the sentence we began studying, are we perhaps justified in setting up the reverse position – as I think we are, because Self and Ego attributes are habitually reversible – namely: 'At the end is number'? As it happens, though this is not especially what I wish to discuss at this precise point, this second statement has a remarkably and uncomfortably prophetic ring about it. One thinks perhaps automatically of nuclear fission and the appalling effects of the scientific revolution on our general environment, both psychological and natural. But leaving actual prophecy out of it, this new statement, in any case, leads us straight to the heart (the head) of the Ego – that is, to number.

In fact, as we in part already saw, words make only poor scientific tools. In that context they are clumsy, inaccurate and full of redundancy. Though scientists have been, and are, at pains to create objective vocabularies, and though, for instance. the employment of Latin roots in English speech, as opposed to Anglo-Saxon roots, does allow us to achieve rather more distance from any subject matter, in fact words do not and cannot do anything like the job required.[11] The scientific revolution *required* number, and it is mathematics, not words, that we must thank for our material and industrial progress and our control of the physical environment. Even ordinary numbers are less suitable – for instance, for computers – than are notional

---

[11] The Ego, then, uses words as numbers. The Self treats numbers as words – in numerology, magical practice and so on. (Similarly, the Ego treats people as objects, the Self objects as people.)

numbers ('numberness', as it were), such as the binary system employs.[12]

If, as I have suggested, words and ideas are in the first place organisms, and idea-systems are species, is it possible to duplicate the phylogenetic tables which are a regular feature of books on evolution? I first made this attempt taking as my two 'kingdoms' (the equivalents of the vegetable and animal kingdoms of evolutionary description) the 'arts' and 'sciences'. The table did not progress very far. In trying to place psychology, for example, I could not wholly justify its existence in either kingdom – and indeed to place it at all was obliged to assume interbreeding between animals and vegetables. Actually the table began more to resemble the family tree of two interrelated royal houses. That, as it happens, is not wholly unacceptable – for family trees are a valid organic-evolutionary phenomenon. Matters improved when I replaced the kingdoms of arts and science by the kingdoms of words and numbers. Now psychology-as-words (e.g. psycho analysis) could readily be separated from psychology-as-numbers (e.g. learning theory). Of course the 'territories' and 'food consumed' of the two species overlapped considerably. This is not of importance – I thought for example of the parallel situation of bacteria and moulds both feeding and living on the same piece of

---

[12] I do not believe, however, that even in the purest forms of number the role of the Self diminishes to nothing – any more than I would conversely claim that the Ego has no meaningful access to words. I am not now thinking of such matters as the fact that the word 'five' and the word 'finger' are generally agreed to be from the same root (and as I personally hold, though I must stress that the view does not have the direct support of etymologists, that 'toe' and 'ten' are also cognate – cf. Old High German *zeha* = toe and *zehan* = ten; and the corresponding Greek roots *dek* and *deka*, further cognate of course with 'digit'), for, like children today, man originally counted on his hands. I am thinking here rather of the fact that the decimal and metric systems – which are quite genuinely and objectively much easier to handle than other counting and measuring systems currently in use – are based on ten, and that we actually *do* have ten fingers. Is this a quite meaningless coincidence? I believe that the men who evolved the *I Ching* might have been able to tell us. (Of course, they did rather take the view that there is no such thing as coincidence.) This power of viewing with the Self we ourselves have so much lost – numerology being, I suggest, a kind of quack memory of what was once appreciated.

Those same men would, I think, have approved of the inversion 'in the end is number', and would not have been surprised at its accuracy on so many levels: at the end is science; at the end are too many people; in the last analysis the ultimate components of matter are mathematical equations.

meat: perhaps digesting different parts and forming different excreta, but still ingesting the same material.

Though I am not reproducing here any of my attempts at evolutionary tables, for they were far from perfect, I found and find certain notions satisfying and more than merely figurative. (1) The notion of words giving rise to many species and genera – (primitive) 'religion', 'philosophy', 'politics' and so on. That is, I think the very first words referred *subjectively* to *inner* states, rather than to objects. (2) The notion of words evolving to become somewhat less word-like – i.e. more objective and referring not to inner states but external phenomena – so that the names of the fingers literally first become the names of numbers and finally (and gradually) autonomous numbers, now conceived of (a) quite separately from the hands which gave them birth and (b), later still, independently of the external objects which at first stood for the numbers (cf. Latin *calculare* = to count with small stones; or the use of the abacus). And (3) the notion of these numbers then invading areas formerly the province of words (as, say, mammals descended from, and triumphed over, reptiles), wresting these from their own former ancestors – but, unlike the mammals, *unable* to crowd words into extinction (though they are still trying hard).[13]

---

[13] In all this (*this* included at one stage attempting to consider words as enzymes) I have found my attention turning continually to the various pairs of 'kingdoms' – organic/inorganic; vegetable/animal; female/male; arts/science, and so on. And I now wonder whether these divisions are all (and merely) projections of the Self and Ego, or whether any are somehow, independently of human beings, in some definable sense real – unlike the man-made divisions between, say, chemistry and physics.

My basic view, and this I stressed very much in *Total Man*, is that we can never know what anything is like – indeed the notion is ridiculous: for all perception involves a means of perception, and the means of perception always to some extent decides what is seen – that is, what the properties are of what is viewed. I am inclined now to feel that it may be possible to have some idea in some cases of what things really are in their own terms, in a sense I shall describe. To put this another way, if any effect of any 'force' at all on the 'resistance' studied remains in some definable sense a constant, perhaps we may take that characteristic of the resistor to be real or actual – because the same – no matter what object or means we use to view it. Let us consider here size. Viewing with our eyes we judge one of two objects larger than another. We check this finding by some objective means and discover that the objective measures confirm our subjective impression. In objective measuring what we do in a sense is to cause one object to 'examine' another (in terms of a third object, the actual measuring device, of course), with ourselves as bystanders. If now *every* object we can bring to bear on the problem (be this passage of light across the bodies, the gravitational attraction beween them, or whatever) 'tells' the same

Symbolic Evolution

My hope in this chapter is that in general I have succeeded in demonstrating at least the possibility of thinking about certain phenomena meaningfully in terms other than those of objectivity and the scientific method. The scientific method is, after all, only a way of looking at phenomena – and nothing more than that. It is at no time *more* than a human behaviour. To say on the other hand that I have *succeeded* in speaking from the standpoint of the Self would be claiming far too much. We as a species have so very much lost sight of the Self (except, perhaps, in art) that we understand almost nothing of its true nature. One glance at the *I Ching* shows this sufficiently. How would one *begin* to put this together from scratch?

True Self thinking is 'magical' – not magic as the Ego misunderstands it. Magic-from-within, not magic-from-without. In what we *normally* refer to as magic, the Ego is attempting either to mimic, or to force, the Self-reaction. The first involves endless misunderstanding, while the second merely causes the Self to withdraw from us still further. Consider, briefly, the magic spell. The uttering of certain words or pseudo-words is supposed (by the intellect, that is) to produce effects in the physical universe – e.g. to turn objects into gold. But, as I conceive it, the 'magic spells' of, say, Neanderthal were just (just!) the discovery that with words, the right words, you can reach into another human being; that the wall of silence which stood (and stands) between beast and beast from the beginning of all life was breachable; that indeed love can be in two hearts at once or one thought in a thousand minds. So, there, in the word (the expressed idea, of course) was the beginning of humanity.

There is of course much more still to Self-thought than this. The *I Ching* is, I consider, truly 'magical' in something of the mystical, though not mystifying, sense of that term.

I have in the present chapter, then, tried to indicate as one of my

---

story, can we say that in the material world the one object *really is* larger than the other?

Scientists may feel that all I am proposing here is what science already in any case does – and that my whole argument against the objective method is thereby demolished. I think not. Hardness (permeability), for example, would not remain constant, even in the relative sense I have suggested, under different approaches. Light passes through glass, but not fog: while a bird can fly through fog, but not of course glass.

I seek to argue this point mainly because I wonder whether the difference between inorganic and organic matter is a real one – or whether there is in fact a simple continuum of events between the two phenomena – one that we fail *readily* to perceive because of one kind or another of anthropomorphic involvement in the issue and/or some simple inadequacy in our terms of reference. This is a complex issue we shall need to consider again.

aims something of the sometimes appalling limitations of the scientific method, if that *alone* is to be our mentor. My more important aim, for which we have partly prepared the ground, is, however, to begin to make a basis for the concept of 'symbolic evolution'. For this we return once again to the statement: 'In the beginning was the word.'

I believe that this sentence refers to and describes the *consequence* of the emergence of language in proto-man.[14] I believe that the compiler or compilers of St John's Gospel understood very well the significance of language as a truly giant step forward in man's evolution. The moment of language was indeed for man, as I have already suggested, *the* beginning of just about everything that we now think of as human. If we look today at the chimpanzee as described by Jane Goodall, we see him as it were shut just outside the gates of humanity, outside so to speak the gates of the universe, the key to that gate (language) missing. It is all he lacks to become human.

Yet there is still more to our gospel opening. I believe it is an appreciation, too, that a new arena of *evolution* has been entered – that of self-controlling (not Self-controlling), as against other-controlled, evolution. From the first living cell up to the higher primates, it is evolution that has been in charge of the organism. With language, the reverse process begins. It is with the word, and its later descendant, number, that man has, effectively, destroyed evolution in the Darwinian sense. Gradually, but ever more surely, the slaves have become the masters, and the former master the slave. Gunther Anders has rightly said: 'We live in the time of the end of times.' Only one time is left, the last of the times – our time.

In more prosaic words, what I am now saying is that *we* very largely control the physical environment which has till this point moulded all life-forms.[15] Already when man first harnessed fire he struck a decisive blow for his freedom and against the environment. A cold winter no longer necessarily killed the physically weak, though perhaps intelligent, individual. Aided by fire, he could wait out the winter. More broadly, man was now self-helped and in turn

---

[14] Unlike the *I Ching*, which in describing and interpreting the movement and change of life offers instruction in the conduct of life – is really, then, a *psychology* of life – I consider parts of the Bible to be fragments of a once complete description of evolutionary processes: a *biology* of life. I have, for example, suggested that the story of Cain and Abel is an account of the evolution of the B_2 dominant from the general ranks of the B_1 dominants. Adam and Eve, too, from one point of view are the story of the two sexes reaching puberty. Eve, correctly, reaches puberty first.

[15] As yet, of course, we do not control the germ cell itself.

was self-helper in all such physical problems; and, with the word, not least in the private sorrows that he could increasingly identify and discuss.

Old-style evolution, in a certain sense, for the moment continues. The change is no overnight one. Ways of thought and ways of life – the 'good' idea-system – increasingly, however, aided and aid survival. Religious belief, for example, might help a man through a crisis that would otherwise kill him. Words of encouragement and cheer produce fresh efforts in the weariness of battle or the course of illness.

Still, of course, man fights and kills man. There is nothing symbolic about this. A war today is still not much different from beast against beast – except that it proceeds without even the checks against it that exist in the wild. But it is now *possible* at least for the conflict to be wholly verbalized. The battle *can* be fought entirely in words and often is (even though, distressingly frequently, physical attack still finally intervenes). There is no longer any necessity for further physical confrontation, either in fighting or in any other area of life. It *can* all take place symbolically. At least in principle, today idea competes with idea, and it is the weaker idea, not the weaker person, that perishes. Or a problem or a hazardous situation is attacked notionally, until solved. No actual physical risk to persons *need* be run. Our *notions* can go where once we ourselves had to. Other animals have not this facility.

In these various senses, then, we remove ourselves from nature's front line – from the area where the moulding of species has always gone on. In this sense, let alone in what other, the word, and the ideas that words clothe, take on, and steadily begin the defeat of, the universe. The holders of the word become inceasingly the masters of that universe; the defeat began, actually, with the first word that voiced the first thought.

V

THE INWARD UNIVERSE

7
The Nature of Consciousness

Whenever I am asked – whether in a spirit of criticism or otherwise – to describe or define consciousneses, my first impulse is always to say, 'But you know what it is'.

The implication behind that reaction is that consciousness is (to use R. D. Laing's words) primary and self-validating :[1] and that all other facts and considerations of mental life proceed from *it*, and from no other point. Descartes might perhaps better have said, 'I have consciousness, therefore I am.' For without consciousness there is no existence. Or rather, perhaps there is existence – that is, perhaps the sun for example *is*, absolutely, with or without any organism present to view it – but there is in that existence no awareness of being, no life. It may be that the point is clarified by creating a distinction between existence and being. The sun has (we assume) some independent existence of its own. It exists. Man also exists and has existence in that sense. But he further has *knowledge* of his existence. He is *aware* that he exists, he is *self*-aware. Self-awareness of existence is, perhaps, being.

My more scientific colleagues might perhaps go along with this. But (I think they would insist) consciousness is not a *thing*. That is to say, they would not accord consciousness the status (!) of an object. Since the study of objects (objectivity), they would say, is the business of science, therefore whether or not consciousness exists or otherwise *is*, in some definable sense, the scientist – meaning here especially the scientific psychologist – may ignore it and proceed about his business.

If my imaginary critics have been willing to go along this far, then I have a definition of consciousness which they might also

---

[1] *The Politics of Experience* (Penguin, Harmondsworth, 1967).

accept – one which I am most reluctant to abandon myself, even though I no longer consider that it adequately describes the full position : namely, that consciousness is functionally dependent on but experientially autonomous of the nervous system.[2]

This definition permits of a 'Caesar and not Caesar' solution. That is, if consciousness is entirely functionally dependent, and since the scientific psychologist is only concerned with function, he may then go his way and I mine without further dispute. Whatever I go on to say about experience or being, as defined, he will allow me to say, for in any case he will not be listening. On my own definition after all, the 'shape of consciousness' is determined by the 'shape of function' – not the other way about.

I am sorry (also for myself!) not to be able to rest content with some such amicable and neat arrangement. For I am unable to escape the possibility that consciousness is not the by-product, as my definition implies, but the central product of the evolutionary process.[3]

There are numerous ways in which one may begin to question the notion of consciousness as a by-product. One line of approach might be to attempt to show an increase in consciousness in organisms as the phylogenetic scale is ascended – an attempt we shall in fact make. Another would be to show consciousness as continually adding to its potential – that is, to seem to be developing functions of its own aside from and other than those of the observable physical nervous system. For this would of necessity contradict the essential nature and the definition of a by-product, namely a secondary or accidental product or result, a subsidiary article produced incidentally. I think we shall indeed be able to indicate attributes of consciousness which are hard to accommodate under the heading of either by-products or accidents; and it will become increasingly harder, I hope, for us then not to see consciousness as the central actor (too often, certainly, the small pathetic puppet) on a very, very elaborate stage.

---

[2] Perhaps I may remind the reader here of the analogy between the nervous system and consciousness suggested in the introductory section – that is, that consciousness perhaps somehow forms around the nervous system in something of the way a magnetic field forms around a wire conducting an electric current. The field is not the current, even though called into existence by it and wholly dependent on it.

[3] 'Product' begins to sound suspiciously like 'purpose'. The suspicion is not unfounded. 'That way madness lies,' as Lear said in one context and a scientist might echo here. Personally, I feel that madness lies in the other direction, as I have often enough indicated. However, if the charge here is one of attempting to introduce the thin end of a 'why?' question into what, according to science, is exclusively a 'how?' framework, then I am afraid I have to plead guilty.

I would like first to examine here an incident which occurred to me while I was a school-boy. It is a commonplace enough one, and what I experienced has been experienced by many others — it is simply more convenient to speak of what one knows well.

During a game of Rugby football out of school hours on a Saturday morning, I must have been kicked or otherwise knocked on the head, for, though I have no memory at all of that incident, I came to to find myself standing on the touchline. I had no idea who I was, where I was, what I was, or indeed any attitude towards, or any memory of, anything at all. I remember, however, that as I watched the creatures or shapes in front of me for a while (this was the other boys continuing the game) I was aware of being puzzled, as if there was something I ought to do or something I ought to recall. Then a further period of blankness ensued and I next remember being outside the school ground, walking down the road looking at the gardens. In fact, in the interim, I had somehow got changed, packed my case and was now on my way home. I remember wondering what roses-bushes were. I later vaguely recall being at a bus-stop. I do not remember getting on a bus, but as a next memory found myself sitting on one. At that stage I had a feeling something was rather wrong, and that there was something I should try to remember. I felt I ought to have a name. I also vaguely felt that something or other had happened to me, and wondered what it was.

The account of the incident this far will suffice. In fact I reached home, and throughout the rest of the day kept remembering things about myself. It was rather like putting together a jig-saw puzzle, and as in jig-saw puzzles towards the end the process went rather rapidly. I was never able, however, to recollect any details of the actual accident.

That original coming-to was, I suggest, fairly close to pure being, or what we might call unmediated consciousness. Certainly my eyes were recording sense impressions and I was aware of that (not, of course, that I had eyes or that these were sense impressions), but at the same time I had no labels or referents whatsoever for those impressions — nor was I aware that they could have labels; nor, really, was I aware of any sense of 'they' either inside or outside of me. What the experience to me very much demonstrates, I think, is that *the contents of consciousness* and *consciousness itself* are not one and the same. Just as in the adult thought and language appear to be the same event (so smooth and instant is their dialogue) so as a rule as adults do we readily regard consciousness and its contents as the same, as one indivisible phenomenon. In cases of severe amnesia, but not only in such cases, we appear to have some evidence that this is indeed not so.

As further evidence we may take the again commonplace reaction reported by individuals involved in car accidents, aeroplane crashes, and so on, in the moments before impact, or sometimes too after the death of a loved one. Some people report under such conditions a marked and unusual degree of detachedness from events ('this isn't happening', 'this is happening to someone else', and so forth), sometimes a marked slowing-down of time almost to snail's pace, and/or a complete absence of fear, with time to reflect (objectively, of course, only a few seconds) in some detail about past events or aspects of their own character – the origin, possibly, of the expression 'his whole life flashed before his eyes'.[4]

Orthodox psychology faces a difficult task in attempting to accommodate these events and the mental states accompanying them, if they are as described. In general, I regret to say, psychologists solve this problem by taking no notice of it.

I personally find my explanation (or at any rate *look* for an explanation) of the detachment experienced in amnesia and in moments of crisis in terms of the nature of consciousness – of which more shortly. But let us for the moment return to the details of the amnesia incident. Although I had, and have, no recollection of those events, I *did* return to the changing room, I *did* get changed, pack my things and walk to the bus stop. I *did* get on the right bus and off at my home stop.

Now, if I did not consciously issue these orders to myself, then who did? For an answer we may for once, I think, legitimately turn to the computer for an analogy. The organism without, or in the absence of, consciousness functions *robotically*. It functions then on the lines along which it has been programmed (a) by inherited-instinctive behaviour patterns and (b) by experience and conditioning. Though the tasks performed by my 'robotic' nervous system were not at all simple, it was very familiar with (i.e. had been well programmed or conditioned in respect of) all of them. The performance was not so very much more remarkable than that of a modern jet airliner flying on automatic pilot. That machine flies a set course and can make corrections within limits to meet changed conditions – i.e. can 'deal with' novel situations, *provided they are not too novel*. What, as I believe, neither the robotic organism nor, of course, the automatic pilot can do is to perform acts outside their

---

[4] On occasion, because apparently time has subjectively slowed to such a marked extent, an individual is able to perform complex or rapid actions which he or she would certainly not achieve in the same time under normal conditions. I have listened to first-hand accounts of such incidents, in some cases corroborated by independent eye-witness, and am inclined to give them credence.

range of previous 'experience' or in any case beyond their inherent capacities.

The topic of learning is a particularly vast one, which I do not wish to consider in too much detail. I suggest, however, that we must distinguish between 'robotic learning' (which does not at all necessarily mean unsubtle learning) and learning via conscious experience, conscious decisions or as the result of deliberation. What I term 'robotic learning' is probably pretty much identical with conditioning (both operant and classical). The circumstances under which this learning (or recording) takes place have been often and amply described by the conditioning theorist – and have to do with schedules of reinforcement, rewards, stimulus differentiation, and so on. This process of conditioning goes on more or less continuously throughout our lives. We may or may not realize that conditioning is taking place – but even when we realize it, or become aware of the outcome of it subsequently, we do *not* experience at any time the actual underlying processes, the actual laying down of the traces. (Thus, when I have lived in a strange district for a time, I realize that I have become familiar with it as a general background – but I have *not* observed that actual conditioning as it has taken place somewhere within my nervous system.) With conscious learning, on the other hand, the case is otherwise. Before proceeding to it, however, let me stress that very often robotic conditioning of the material that I am learning or studying consciously is taking place at the same time.[5] Thus some learning – perhaps even most – is stored in two ways at once. There are a number of differences between *precisely* what is stored in the two cases – robotic learning tends to be more indiscriminate, for example, or at least to be discriminative in terms of its own – but this need not concern us at the moment.

To approach the subject of conscious learning, a short digression is necessary. In a course of classes on meditation which I once attended, one of the exercises we practised at home was that of observing ourselves, but of observing without interfering. Thus the instruction for one particular week was 'observe yourself walking'. Now, as we all know, normally if one starts thinking about the mechanics of some task which one has learned to do without thought – such as typing, driving a car, walking upstairs – the result is that that smooth activity is disrupted, and we make mistakes. This was not at all the purpose of this exercise. The purpose was to observe *without* disrupting. Some of us found this very difficult at first, but

---

[5] Experiment has shown for example that a memorized task is better recalled in the room where the original learning took place than in an otherwise generally similar room.

with practice we all managed it. (This ability to detach oneself from and to observe what one is doing without in any way reducing the level or rate of performance is already of considerable interest to our general argument in that it recalls the detachment sometimes experienced during accidents.) The result was not at all unpleasant or tedious, but on the contrary both interesting and rewarding. This exercise represents one small example of the 'rediscovery of the senses' which many programmes of meditation incorporate. Another similar exercise we practised was 'observing oneself in the house'. In the class preceding that exercise, the instructor asked us to suggest examples of things we obviously had done or habitually did, but could not remember doing. My own mundane example happened to concern my spare shoes, which I always found under the bed, although I took them off and left them in various parts of the room. I could only imagine that in the course of moving about the room during the day I gradually kicked them under the bed. I had no memory of so doing – though I subsequently observed myself doing it.

Further instances of the foregoing in fact happen regularly to all of us. That is, we cannot remember where we have put something – sometimes we cannot even remember when we had it last. When we find the object eventually, we sometimes retrospectively recall putting it where it now is. In still other cases, however, we do not remember placing it where it is – yet we must have done. For it is, for example, in a locked cupboard to which only we have the key. It is under the tool-bag in the car. It has got into the parcel we sent to Scotland. Like the events surrounding a blow on the head, the incidents in question defy recall – in the present case, I suggest, *because they have never been recorded.*

The last statement requires amplification. The events surrounding the blow on the head may or may not have been recorded. But if the former is true, the actual (physiological) memory trace seems to have been in some way ruptured beyond repair. The shoes kicked under the bed, the book put back in the cupboard, the key included in the parcel are not in this category. It would seem that they were not recorded (a) by consciousness, because our conscious attention was somewhere else at the time, or (b) by robotic memory, because, I suggest, the actions were performed once (and no kind of reward was involved) on each occasion, and as a rule one trial or one reinforcement is insufficient to produce this type of learning.[6]

The differences between conscious learning and conditioned or

---

[6] The kicking of the shoes under the bed is a frequently repeated action, and one would perhaps have expected conditioned learning to have occurred. However, although robotic observation is fairly open-ended (rather like a perceptual vacuum-cleaner) – so that most of what goes on is noticed

robotic learning seem to me many and clear. Another, broader version of essentially the same statement is that behaviour which involves consciousness (conscious participation) is markedly different from behaviour which does not. Supportive evidence for this view can be readily found in any experimental psychology text – though of course the terms used are not those that I employ.

An experiment by Hull involved the use of Chinese characters.[7] Unknown to those concerned in the experiment, a particular sub-character was buried in some of the larger characters. The subjects were asked to guess-sort the cards into classes. Although performance improved on each run, the subjects still did not know at the end why they were getting their choices right. Unconsciously, and robotically, they had been conditioned to make the correct response without knowing why.

The point of interest in this experiment for present purposes is that, if the subjects had simply been told what to look out for at the beginning, without doubt most of them would have achieved 100 per cent. correct sorting from the start. Some experimenters have actually confirmed this in similar experiments, though in fact the point really needs no demonstration. The matter in question is frequently discussed, as a special issue, under the heading of 'one-trial learning'. This title is an attempt to bring this kind of 'learning' (i.e. the instant appreciation of or acting on a conscious instruction) into the general rubric of conditioning or, as I call it, robotic learning. That this is a tortuous, and in my opinion quite unsuccessful, attempt is seen when we consider a few everyday real-life situations. If a friend asks us to go to his house (which we have not been to before), retrieve the key from under the brick by the wall, go into the house and pick up the parcel on the sideboard in the lounge, we can do it, just like that. We have 'learnt' to do it in one trial. I suggest actually that we have 'learnt' to do it *before* the trial itself, but no matter. Conditioning theory, for its part, is obliged to argue that we have been to many houses and many lounges and seen many parcels before (which is true), so that in a sense earlier conditioning *has* occurred; and that the reward or reinforcement in the present

---

and recorded most of the time – it is not infinitely open-ended. Some preselection of stimuli occurs, and some types of event have preference over others (situations connected, no doubt, to our phylogenetic history). Moreover, I believe the robotic view-finder, though very wide-angled in the young child, narrows progressively with increasing age. Probably too the actual laying-down of the robotic memory trace requires more reinforcement in later life. These suggestions are in general in line with the evidence obtained from conditioning experiments.

[7] C. L. Hull, *Quantitative aspects of the evolution of concepts* (Psychological Monographs, No. 123, 1920).

case is that of retaining the affection of our friend. This already laboured analysis becomes even more unworkable if we consider the next situation. We are on a train, and a total stranger sitting in the compartment requests: 'Would you mind saying "Smith" to me when we get to the terminus – I must remember to post a letter before I leave the station?' Where is the reward here, and where the previous trials? For some of us, the strict conditioning explanation becomes farcical at this point.

Yet another and even more challenging difficulty resides for the conditionist in so-termed 'latent learning'. Let us suppose that for the first time in one's life one is asked by an acquaintance where he can purchase some ice-skates (a Tibetan rug, a glass vat for storing acid – or whatever the recherché item is). We have never thought of the matter before. But now, on reflecting, we recall that in a part of town we occasionally visit there is a sports-gear shop with some ice-skates hanging somewhere in the display. All we have ever done is to walk past this shop, perhaps only once in our lives, as with hundreds of other shops, paying no particular attention to any of them. Yet when needed – and we can never have known that information would be needed – we can readily recall it. Does it not seem, then, as if consciousness is in the recording business on its own account, independently of any material or other reward or advantage, just somehow as part of the process of being conscious? The same type of incidental, non-rewarded learning can, as it happens, be shown to occur also in lower organisms.

Further light on consciousness, and further difficulties for conditioning theories, come from experiments where subjects variously are and are not told certain things – are given or not given certain instructions.[8] In a series of experiments one experimenter tried to produce a conditioned salivary response in human beings to the sight of nonsense syllables flashed on a screen, via the eating of pretzels (the unconditioned stimulus). The results were chaotic. Those who saw through the experiment developed no salivatory response. Some others produced the response for a while and then lost it, others produced a *decrease* in salivation, and so on. Eventually the experimenter hit upon the idea of misleading the subjects. He fed their consciousness false information as to the nature and purpose of the experiment. He told them the experiment was about something else altogether. Under these circumstances he was able to produce normal Pavlovian conditioned salivation in all his volunteers.

As Woodworth and Schlosberg point out, one is faced with the

---

[8] See R. S. Woodworth and H. Schlosberg, *Experimental Psychology* (Methuen, London, 1955), pp. 572-6.

problem of how conscious attitudes can control or affect a response like salivation, which is involuntary. The mystery is deepened when the same general picture emerges from experiments involving responses which people *do not even know they possess,* like the psycho-galvanic skin reflex. Apart from the difficulties which these experiments produce for the conditioning theorist, we see here very much that consciousness (conscious attitudes) has at times a surprising degree of influence and actual control over the robotic or involuntary reactive systems.

I would like to suggest that learning via consciousness is otherwise, and is other-where stored, than is robotic learning – be this of the classical or the operant variety. Under conscious learning we must, of course, include everything that is learned by study or rote or deliberation or cogitation – the method for solving quadratic equations, the history of the eighteenth century, the speeches of a character in a play, or whatever. I suggest, too, that anything that has been learned consciously – and here we must broaden the rather narrow term 'learning' to include *all that which passes through conscious awareness in any form at any time,* so therefore daydreams, and all thoughts of every kind – is stored and is potentially available to consciousness again; is stamped, as it were, 'may be wanted on the voyage'. While, on the contrary, neither (a) the *processes* of robotic learning as such nor (b) *necessarily* even the *results* of robotic learning are available to consciousness. (As a rule, however, the results – effects – of such learning can become available, though certainly not always, once attention is directed to them.)

Thus the subjects of Hull's experiment had robotically learned, or partially learned, how to solve the problem of the Chinese characters. Whether or not they consciously knew this (actually they did not) was of no importance. Being told the truth subsequently (though this, of course, caused further immediate conscious learning to take place) did not bring the *conditioning process* itself into consciousness (only the results of it), nor could the conditioned learning be thereby unlearned, though it could then, at least in principle, be inhibited or rejected – i.e. not acted upon – by their consciousness on subsequent occasions.[9]

At this apposite point let us also note the undoubted ability of consciousness to choose to reactivate past memories. I can *decide* to

---

[9] The fact that consciousness can choose to do the 'wrong' thing – go against the results of conditioning and/or the known correct answer, here admittedly more decisively in matters of operant than of classical conditioning – again very much suggests that consciousness has powers of action independent of the jurisdiction of those conditioning and reinforcement schedules.

remember what I did two weeks ago last Thursday. One at this point, so to speak, goes through the files and gradually or otherwise recovers some or all of the appropriate memories. Do we see that these memories and consciousness are not the same thing? They – the memory traces – are capable of being reactivated so that they once again become part of consciousness. But they are not *in* consciousness or *part* of consciousness while they are memory traces in storage. Again we can argue that consciousness and the contents of consciousness are not the same.

To digress briefly: it will be realized that we begin to need some fairly sophisticated designing at the physiological level to meet even the proposals I have made so far. We need (1) receiving, processing and storage centres for classical conditioning, (2) receiving, processing and storage centres for operant conditioning, and (3) a generally similar arrangement for material passing through consciousness. Let us now make it still more complex. We need yet a further centre for material passing through consciousness *when that consciousness is inhabiting the Self and not the Ego* – that is, notably, when we are dreaming.

I have said a good deal on the subject of dreaming in *Total Man*, and do not wish to repeat all arguments here. But let us briefly run through the following points. For something, I suggest, in excess of 99 per cent. of the time we are dreaming, we have no memory of our waking/conscious lives (that is, in the terms of reference of that consciousness), i.e. no knowledge of waking consciousness and no understanding that we are dreaming. At the time of dreaming, dreaming *is* our life, and there is no other. Secondly, on waking we recall, as a rule, only the merest fragments of what we have dreamt (even though the normal adult in fact dreams for upwards of two hours a night) and unless we proceed to rehearse this by thinking it over consciously while lying there, these fragments too will disappear in the course of minutes. And even so, our conscious rehearsal of the dream is, in any case, not the *same* as the dream – not only not in its texture and feel, but also in the sense that we, inevitably, embroider and edit the material. From both sides – from waking consciousness and from dreaming – there is then more than a suggestion that some alternative location of activity is involved in the two cases. It is of course true that much of the material of the dream consists of fragments of information and situations from waking life. These fragments, however, are used and experienced in quite different ways, so that they are very much novel items, even if they reveal traces of their origin. As I have proposed, it seems that Self-consciousness can 'raid' the memory store of waking consciousness during the latter's absence, and borrow what it needs for its own use. Aside from *this* material, there seems other material in dreams (in nightmares, for example, but also in revelatory dreams) that is Self-generated – that is, does not derive from waking life.

Having added centres for the receiving, processing and storing of Self-conscious material, we have still not finished. I have said that material which has once passed through consciousness and been stored can always potentially be reactivated and made conscious once again. The sometimes astonishing extent to which this appears to be true can be demonstrated both under hypnosis and during psychoanalysis where (sometimes objectively verifiable) data are recovered from perceptions in the very earliest months of life. All this concerns the present life of the individual. To my statement I would now like to add the following: that certain material *which passed through consciousness* in our ancestors, including very distant ancestors, can also potentially be reactivated into present consciousness. This, effectively, is Jung's 'racial memory' or 'collective unconscious'.[10]

What we are discussing here is *not* the personal, day-to-day experience of our ancestors. There is no suggestion of reincarnation. But in ways we do not readily understand it seems that some experiences – especially I think archetypal experiences – become genetically encoded in a particular generation, and henceforth become the property of all descendants. Not only do we recapitulate something of these ancestral experiences in the course of our growing up (consider how much more powerful the hero archetype is in most young boys than in most adult males, for example), but we can under some conditions in some sense go back into them. To stay with the example just given, despite our, as adults, habitual dismissal, or simple no-longer-awareness, of the hero figure and the heroic situation, a certain story, a military march, a particular incident may suddenly set our heart beating faster, create a momentary re-vision of heroism in our mind, a vision of *oneself* as hero fighting back uncounted odds . . . ! But then we pull ourselves together.

One further complication: just as Self-consciousness can raid the memory store of waking consciousness, so I believe it can also raid the *race memories*, that is, the archetypal contents, of the drowsing Ego. And finally, there is the probability that the Self has archetypal memories of its own.

Indeed, it seems we need a dauntingly complex system to meet even the superficial workings of the human psychological organism. How shall we hope to find anything remotely resembling this in a few million blips on a few miles of magnetic tape, or the allied

---

[10] Some individuals maintain that under such drugs as L.S.D. they achieve 'memories' of pre-human states. I do not wish to affirm or refute such claims here, but only to note them.

models of the academic psychologist, that would be merely laughable were they not so tragic.

In speaking of the reactivation or recovery of memory traces by consciousness – especially perhaps when not single memories are involved but, as in the case of amnesia, very large areas of experience – there is perhaps some sense of *moving into* those memories. This, I would choose to say, is the case. Consciousness, whatever it may be, seems subjectively to be capable of somehow shifting its locus in the nervous system. This possibility is already suggested by the finding that, during cognitive thought, problem solving and similar activities, a certain type of electrical activity is recorded from the front of the brain (beta and gamma waves); during daydreaming, especially with the eyes closed, and in states of meditation, a quite different rhythm is detected primarily at the rear of the brain (alpha waves). I base my own main case, however, on rather different 'evidence'. We note firstly, on the one hand, the clear reference to spatial movement in the very large majority of our metaphors relating to mental states of all kinds : out of my mind, off my head, beside myself, falling in love, falling asleep, feeling 'low', feeling 'high', down-to-earth, way out, flights of fancy, walking on air, floating on a cloud of dreams, on top of the world, absent-minded, withdrawn, and so on. There is further the central role of the archetypes of the Journey, the Wanderer and the Pilgrim in many legends and stories of both past and present, including frequently, and still at the present day, in folk-song. In particular, going (we even say 'going') to sleep is very often described, especially to young children, as a journey – to the Land of Nod, the Land of Dreams. Such metaphors, I have proposed, are archestructural perceptions of the movements of consciousness within the nervous system.

All in all, then, as must by now be very apparent, I find myself obliged to consider consciousness as some kind of entity – not merely some aspect of conscious contents, and non-divisible from those contents : therefore *un*like, say, the colour of a car, which, although *conceptually* abstractable from the car itself, cannot and does not exist without the car (or some other object) – and indeed is nothing more than an aspect or attribute. My earlier definition of consciousness – functionally dependent though experientially autonomous – will therefore not suffice. For it seems that consciousness, at any rate in human beings, has definite functions and powers of its own. Nor have we, by any means, yet said the last word on this score.

Consciousness, I suggest, is more closely associated with recently evolved parts of the nervous system than it is with older parts.[11] We

---

[11] Thus we would expect Ego-consciousness to be stronger and more definite than Self-consciousness – and this seems to be the case if we com-

may think up to a point in terms of a coral bank. Coral flourishes best just below the surface of the sea. Where the sea-bed is gradually sinking, as in parts of the Pacific, the lower coral dies, while the still living coral builds further upwards towards the receding light, on the myriad skeletons of its own ancestors. Hence arise coral reefs and coral banks. We may liken consciousness to the few feet of living coral atop perhaps hundreds of feet of dead coral. The analogy is by no means perfect, for I do not suggest that the lower levels of the brain are dead: they are of course alive, but merely, for the most part, not illumined by consciousness. Yet the facts of the coral's life, plus the further fact that its textural structure is not dissimilar in immediate appearance to the material of the brain, plus the yet further fact that the sea has in its own right any number of associations with the unconscious – these, I think, are the explanation of our great interest in coral. I believe that our perception of it is in fact archestructural. And one species, indeed, is actually called brain-coral.

A further as I suggest archestructural perception of consciousness involves the firefly or Will-o'-the-wisp. This tiny point of brilliant light moving in a void of surrounding darkness, dipping and rising, alternately flaring and waning or lost to view, very much reminds us again, I think, of the behaviour of consciousness in the 'blackness' of the inactivated cortex. As also a man might carry a flickering lantern down endless rows of subterranean files, choosing one here and there to illumine momentarily.

On rather another tack, it is a very long time since I first puzzled over the reflexive verb, without being quite able to understand why I found it so puzzling and so intriguing. I think I now do understand why, and I believe that in this phenomenon we gain yet further insight into the nature of consciousness.

In the sentence 'I wash myself,' who is the 'I' and who is being washed? (This, essentially, was the problem over which I puzzled.) Now, it is certainly possible to argue that all we have here is a semantic confusion. The sentence really means 'I wash my body.' Only one entity is involved. Well enough, but what then of the sentence 'I hate myself'? Does this mean 'I hate my body'? Not

---

pare consciousness while waking to consciousness during dreams. Consciousness is also enlarged by (takes its quality from) the (sensory) equipment it has available. Thus waking consciousness has full colour, while the large majority of dreams are in monochrome – shades of black and white. Monochromatic vision is phylogenetically very ancient, colour vision being of fairly recent origin. The Self, as I suggest, for the most part makes use of visual centres designed to receive and interpret black-and-white vision. These centres were originally partly connected, I further suggest, with the pineal eye of the pre-mammalian organism.

necessarily. A person may love his body, but hate, say, his greed and his unreliability. Or what of 'I hate the fact that I hate myself' – which may be rather better expressed as : 'I cannot bear the fact that I hate myself' (that I cannot accept myself). These situations are far removed from and far harder to explain away than the apparently unremarkable 'I wash myself.' If someone says he hates his hatefulness, does he (or can he) hate the hate which he is employing at the same time in hating with it? 'Hates his hatefulness' can of course also mean that he hates the fact that others hate him. Both statements, however, come close to saying 'I hate my personality.' Is 'I', then, not part of the personality? Is 'hate', then, not part of the personality, at least at the moment in question?

In all these extended examples do we not see that the 'I' can apparently borrow attributes which belong to the personality (which are part of the personality) for momentary use against the remainder of the personality – or the overall personality *structure*. If someone says he likes his hatred of evil, he approves at that moment, in that context, of hatred *per se*. But if he says he hates his hatred of his mother, he disapproves of hatred, at that moment. So at one moment he praises this thing (hatred), at another he dislikes it, at a third momentarily he actually *is* it (e.g. I hate my parents). 'I', it seems, is any of these things and all of these things, and none of these things. Actually, I believe the last of the three statements to be correct. The first and second appear to be true – and, in a temporary sense, *are* true – whenever the 'I' 'clothes' itself in a particular attribute or attributes – or, we can again say, *activates* them.

The analogy of a person and his or her wardrobe is, however, a very useful one here. The 'I', it appears, is forever dressing itself up. Just as a person looks different (even feels different) in different styles of dress, so the 'I' seems different when it is loving or hating – and in a way *is* different. But there seems behind every state and every attitude an 'I' which is a continuing entity. (In *waking* consciousness at least, I know that I am the same person who hated last night, loved the night before and was indifferent the night before that.) But like the invisible man, perhaps the 'I' can only be seen (by *others* – be experienced by them) when it is wearing something. *I*, of course, experience my 'I' (*am* my 'I') even when totally detached from all guises. In states of amnesia, in moments of extreme crisis, or when waking alone at night, we become perhaps particularly aware of the 'I' that we are without mediation and, virtually, without stimulation. (If consciousness were merely the outcome of sensory stimulation, it would be paradoxical that in this last case – when alone in bed, with stimulation at a minimum – we are sometimes most conscious of ourselves and the fact of our existence.)

Actors interest us more than many figures, I think, because they represent, archestructurally, the 'I' which over and over again dresses up in the trappings of the personality. They show us our condition *in extremis*. Frequently, off-stage, actors seem to be naked scraps of consciousness, not finding in their own mental wardrobe enough to put on – or perhaps (turning my metaphor round) *over*dressing to cover up a too-sensitive 'I' that fears the glare of reality (daylight and the searching scrutiny) with uncommon desperation. Need I draw attention here to the many metaphors of assuming masks of which we make such play (N.B.!) in everyday speech? To the mask which most women literally put upon their faces most of the time? To the social roles within and behind which the temporarily safe 'I' sits with fingers crossed and eyes shut tight – manager, lorry-driver, secretary, lecturer, housewife, father?[12]

It will have become clear that consciousness is identical with, and named as, 'I'. It must be firmly stressed that 'I' is not Ego, nor yet Self.[13] Ego and Self are but two very extensive wardrobes, into which the 'I' may dip – well, not quite as it pleases, though the freedom of the 'I' in this respect can be considerable. Often the 'I' is the willing or the unwilling prisoner or shadow of the Ego or Self, or even of one or two dominant aspects of those major *personae*. As to the precise relationship of the 'I' to the Person (to System C), that is a very difficult matter conceptually, which we shall attempt to discuss in the next chapter.

The first of the major attributes of the 'I' is that it knows itself to exist. It knows this directly, inherently and without mediation of any kind.[14] It then comes to know something (a) of the conditions of its existence in terms of the organism in which it is housed, and of which it is in some sense some kind of extension, and (b) of the external universe in which the organism resides. These aspects of its knowing are *mediated*, not inherent or direct. (But still the 'I' then *knows that it knows*. It can so to speak invest the incoming information with itself; it can cause information to pass into awareness – to become conscious.) This is a second major attribute of the 'I'. so consciousness continually extends and augments itself. It is as if it

---

[12] In this general context it will probably be clear that I regard (1) the wearing of clothes as such (other than from the purely functional standpoint of keeping warm) and (2) the 'invisible man' stories and the attraction of the idea of invisibility through the ages as further (archestructural) perceptions and expressions of the nature of consciousness.

[13] The conscious Ego is *haunted* by the (then) robotic Self (ghosts, vampires, and so on); the conscious Self is *persecuted* by the (then) robotic Ego (the machine, Midas, the anti-organism).

[14] Cf. both R. D. Laing's description of awareness and Descartes' 'I think therefore I am.'

builds itself a house, a vast mansion (many mansions?), of converted or transmuted sensory information: this house *becomes* the 'I' in the sense that the 'I' can move within it and know it. But the house, nonetheless, does *not* know *itself*. Memory traces when not activated 'are' not, though they do exist in the objective sense.[15] Robotic memories also 'are' not, though they likewise exist, and in their existence form part of the complex programme of the un-self-conscious nervous system. The functioning of this system (or rather systems) cannot become conscious, though here the *results* may sometimes be observed, and *these* then, of course, consciously recorded in their own right.

I have suggested that the 'I' knows itself, that is, has awareness of its own existence, as an irreducible and immutable characteristic. I have further suggested that any extension of knowledge beyond this initial self-knowledge is mediated by whatever (sensory) mechanisms are gathering and conveying the information. In some sense these mechanisms therefore determine the nature (the quality, the dimensions) of the 'I's' *further* or *mediated* consciousness.

This is readily illustrated if we consider the 'I' (consciousness) in the two cases of occupancy of the Ego and occupancy of the Self. During waking hours the normal individual, at least, has the impression that he is exercising control over his actions, is choosing between alternatives. Certainly *part* of this freedom is illusory, in that for example it can be shown that many people avoid areas of activity and situations where the limited power of their 'I' would be revealed – that is, where programmed (conditioned) reactions of the robotic organism would take over instead. For instance, a man might become a bank clerk because, among other reasons, he would not be able to cope without anxiety with people in some less structured, predictable or 'safe' situation. A nervous teacher might choose – not necessarily completely consciously, of course – to teach adults instead of children. A woman might marry a timid, conventional man (or, naturally, vice versa) in order not to have her own uncertainties challenged – and so on. Thus, to repeat, though there is a fair measure of autonomy for the 'I' in waking consciousness, it is seldom as great as it seems. In the *neurotic* individual we see, of course, an 'I' with a very reduced autonomy. But it is in dreaming sleep that I think we can best personally experience a consciousness (our own consciousness) with a very great deal of its autonomy absent. We speak of 'having a dream', but, as already suggested, it is more correct to say that the dream has us. We are very largely at the mercy of the dream and, I

---

[15] Cf. earlier discussion concerning the policy of the Northern Europe Presbyterian Building Society.

suggest, we know that we are. This extension or quality of consciousness (for it *is* an extension, even though the extension is into an unfree situation) derives from the nature of the Self – which has a yielding or, more extremely, a masochistic nature. Sleeping consciousness is perhaps, too, as a rule less *vivid* than waking consciousness – but not always or absolutely so. In a strong nightmare, for example, our consciousness of events possibly exceeds anything that we experience when awake. On awaking from such a dream it may be still 'in the room with us' for many seconds, and in our thoughts for much of the day. It is not therefore the degree of consciousness which is essentially different in the two cases, but the *type* of consciousness. In Self-consciousness the 'I' has a very different experience of being extended (of being *alive*, that is, instead of just being) than it has in Ego-consciousness. But it *is* nonetheless an experience of being alive, of being somewhere, of being in contact with (mediated) events.

In our own Self-consciousness (that is, during dreaming) I think that we come close to getting some idea of what consciousness is like in and for lower organisms, that is, in animals. I say only *some* idea, for the waking consciousness of animals is, nevertheless, a rudimentary Ego-consciousness. But it is a far more trapped, less-able-to-act consciousness than is our own Ego-awareness. The waking 'I' of the animal is much more like the helpless by-stander that is our own dreaming 'I': the witness who must witness but cannot control. In Roah and the dog Tim, however, we saw two examples of what I consider represent *acts* of consciousness, specifically Ego-consciousness, in animals.

My assumption is that a rudimentary form of awareness is present from the lowest, single-celled creature upwards – be that creature animal or vegetable. It is then, of course, very, very much a governed (not a governing) consciousness. The very small human baby, I suspect, experiences awareness in something of this highly circumscribed, largely undefined way – though for a real comparison I suggest we should consider rather the early human foetus. At some point in the evolutionary scale (and in the ontogenetic recapitulation of the scale) consciousness begins to develop or acquire powers of control. The switch point from controlled to (ever so slightly) controlling consciousness is as difficult to discuss or assess as any of the other momentous switch points – the moment when inorganic becomes organic, for example – and I do not therefore propose to discuss it here. I am sure in my own mind, however, that phylogenetically this switch occurred well before the advent of the primates, let alone of man himself.

In man the controlling power of consciousness has reached its highest point so far. It is not, though, such a very high point. Such marginal powers as consciousness possesses are as a rule readily overridden by the commands and demands of the autonomic system, and by both robotic operant and classical programming and pre-programming. Yet I believe that occasionally consciousness, even though its efforts are frequently rather like a pitchfork of straw against a wave, can throw enough into the balance to tip the scales a particular way. In fact I would like to express this notion even more firmly, and quite illogically, by saying that consciousness at certain moments can even drive the organism beyond that of which it is actually capable. There is a clear paradox here, which I do not intend to do anything about resolving: how, after all, could an organism do more than that which is within its capabilities?

Yet it seems to be on these two notions – (1) that consciousness can always seek to reverse, and sometimes succeeds in reversing, the decisions of the autonomic and the robotic systems and (2) that consciousness can demand, and sometimes get, a greater performance than expectation could possibly, let alone reasonably, demand – that the concept, and fact, of morality is based; and only on one or both of these that it *could* be based. Morality in the true sense is no kind of notion that could apply purely to the robotic systems alone – though I have no doubt that some of our values derive from our precise biological predispositions. True morality must, I think, involve a knowing, striving, choosing entity – one that is, and can be, in some sense more than, and greater than, the sum of its past experience.

Our next chapter also will be very much concerned with the nature of consciousness and morality, and with the further general nature of mental phenomena and contents – more especially with the ways in which a content or set of contents meets and interacts with another.

I am very aware that much of what I have said in this chapter could do with more amplified notation and definition, and that at least as many difficulties are raised as are settled by my explanations and descriptions. Yet I feel that what I have said goes at least some way towards meeting the complexities of *readily observable actuality* in life – and certainly further than do any academic accounts of personality that I have so far come across, always excepting the contributions of Freud and Jung.

8
'On the Aesthetic Education of Man'

(i)

The title of this chapter is taken from that of a book by Friedrich Schiller, *On the Aesthetic Education of Man, in a series of Letters*, originally published in 1795.[1]

Friedrich Schiller requires no recommendation from me. He is, for example, the second man of German literature, outshone only by Goethe (whose close and respected friend he became). He is, with that, a major European figure also. He combined the talents of dramatist, poet, literary theoretician, philosopher and historian. His first drama, *Die Räuber* (written in secret and published at his own expense) is, and was then, considered a landmark in the history of the German theatre. His much later trilogy, *Wallenstein*, has been ranked with Shakespeare's historical dramas. These credentials amply suffice – though I myself prefer to base the worth of an artist rather on the extent to which he inspires other artists. So, for example, Verdi gave two of the dramas operatic form; while the fourth movement of Beethoven's magnificent Ninth Symphony is based on one of Schiller's poems.

In Schiller, therefore, we have, whatever else, a gifted artist in his own right. This is, for me at least, of prime importance in the evaluation of his views on aesthetics and the aesthetic experience – that is, of Schiller as critic.[2] The second point, on which I shall be enlarging,

---

[1] O.U.P., London, 1967. Translated by E. M. Wilkinson and L. A. Willoughby. All quotations are taken from this edition.

[2] This term as used here must not be considered to have any common reference to the individuals generally called critics today. These, despite their own view of their function, and the view of those still less able who endorse their activities, are making quite valueless – and indeed wholly

is that in viewing Schiller's range and degree of achievement we have already grounds for suspecting him to be an AB dominant – one of those individuals equally gifted in the higher functions of Self and Ego. Such individuals – among whom I number Goethe, da Vinci, Nietzsche, Michelangelo and Beethoven (all of whom incidentally were left-handed) – providing they (that is, their consciousness) can ride out the storms of their natures (in the end Nietzsche and Beethoven could not; nor, altogether, could Schiller himself) manifest exceptional degrees of creativity.

As a young man Schiller was heavily involved in the German contribution to the romantic movement, then under way in all parts of Europe – the *Sturm und Drang* (Storm and Stress) movement. *Die Räuber* is, very much, a romantic drama concerned with the question of individual freedom, a concern which Schiller never lost. He was, therefore, heavily involved both emotionally and politically with the actual French Revolution of 1789. His third drama, *Kabale und Liebe*, is an instance of the 'bourgeois tragedy' discussed in our first chapter, concerned, that is, with 'ordinary' people (meaning, however, the middle classes, not the proletariat). All of these items are evidence of the powerful influence of the Self in Schiller's nature.[3] That he later withdrew from direct confrontation with the Establishment, and the unqualified championship of emergent psychological and sociological forces, is due to the influence of his own opposing (System A) endowments – and not, I would argue, to any betrayal of those former feelings, as suggested by those who seem by reason of the narrowness of their own vision to find it far easier to be so whole-heartedly one-sided.[4]

Before we move on to the central ideas of the *Letters*,[5] perhaps two

---

destructive – System A statements about Sysem C phenomena. Theirs is actually no function at all beyond that of spoiler and obscurant: as a mouse, nibbling at grains of corn, might imagine himself to have contributed to the harvest.

[3] The romantic movement, the French Revolution and the rise of Socialism (with all its concomitants – equal rights for women, and so on) are each an aspect of what I have termed the 'Neanderthal backlash'. In all this we see the Self (the B-dominant) gradually freeing itself (after virtual extinction, historically speaking) from the tyranny of the Ego, challenging the latter's right to be sole arbiter of human destiny.

[4] There is no more virtue – that is, no less one-sidedness – in being a champion of the socialist Left than in being a champion of the conservative Right.

[5] Schiller wrote at length on these ideas also on many other occasions – for example in his well-known *Über Naive und Sentimentalische Dichtung* (*On Naïve and Sentimental Poetry*).

examples of the essential modernity and relevance of Schiller's thought should be given. These will, I think, predispose us not to dismiss too lightly the more complex and philosophical main argument as being either irrelevant or, if relevant, then merely rhetorical and of no *practical* import. The first extract is his description of modern society – of 1793, that is. How clearly Schiller saw and expressed what the sociologist of the twentieth century is only now re-expressing.

State and Church, laws and customs [are] now torn asunder; enjoyment [is] divorced from labour, the means from the end, the effort from the reward. Everlastingly chained to a single little fragment of the Whole, man himself develops into nothing but a fragment; everlastingly in his ear the monotonous sound of the wheel that he turns, he never develops the harmony of his being, and instead of putting the stamp of humanity upon his own nature, he becomes nothing more than the imprint of his occupation or of his specialized knowledge . . .

When the community makes his office the measure of the man; when in one of its citizens it prizes nothing but memory, in another a mere tabularizing intelligence, in a third only mechanical skill . . . when, moreover, it insists upon special skills being developed with a degree of intensity which is only commensurate with its readiness to absolve the individual citizen from developing himself in extensity – can we wonder that the remaining aptitudes of the psyche are neglected in order to give undivided attention to the one which will bring honour and profit? . . . It is rarely recommendation in the eyes of the State if a man's powers exceed the tasks he is set, or if the higher needs of the man . . . constitute a rival to the duties of his office . . .

Forced to resort to classification in order to cope with the variety of its citizens, and never to get an impression of humanity except through representation at second hand, the governing section ends up by losing sight of them altogether, confusing their concrete reality with a mere construct of the intellect; while the governed cannot but receive with indifference laws which are scarcely, if at all, directed to them as persons . . . Weary at last of sustaining bonds which the State does so little to facilitate, positive society begins . . . to disintegrate into a state of primitive morality in which public authority has become but one party more to be hated and circumvented . . .

(Sixth Letter)

We see, then, that humanity has never lacked speakers of the truth, only listeners to it. How well Schiller's statement could head any modern Socialist manifesto.

The second extract reflects and anticipates the views of romantics such as Wordsworth ('we murder to dissect', the 'meddling intellect', and so forth) but, a hundred years before Freud, also refers to and describes the results of bringing unconscious contents into consciousness.[6] Finally, how Laingian and (therefore) absolutely modern is the latter part of this extract.

> But it is precisely the technical form, whereby truth is made manifest to the intellect, which veils it again from our feeling. For alas! intellect must first destroy the object of Inner Sense if it would make it its own. Like the analytical chemist, the philosopher can only discover how things are combined by analysing them, only lay bare the workings of spontaneous nature by subjecting them to the torment of his own techniques. In order to lay hold of the fleeting phenomenon, he must first bind it to the fetters of rule, tear its fair body to pieces by reducing it to concepts, and preserve its living spirit in a sorry skeleton of words. Is it any wonder that natural feeling cannot find itself again in such an image, or that in the account of the analytical thinker truth should appear as a paradox? (*First Letter*)

These two extracts are, whatever else, models of clear thought clearly expressed. Yet, once into philosophizing proper, Schiller becomes, in my opinion, guilty of some of the standard bad practices of philosophy. It has been said that psychologists collect facts for which they have no explanations, while philosophers collect explanations for which they have no facts. There is a good deal of truth in both statements. One instance of the latter practice is seen in the philosopher's willingness to generate concepts and terms in the head, and to treat them then as if they existed with the reality of external, physical objects – i.e. to treat them as if they necessarily existed by virtue solely of having been thought of. While I have no wish to get involved, at this point, in too much discussion on the general nature, usefulness or otherwise, of philosophy, and while the reader is fully entitled to disagree with me, I find personally that too often philosophers use the mind as a kaleidoscope, turning it endlessly through never-repeating combinations of concepts, which, startling, diverting, even inspiring though they can be, lead nowhere beyond the kaleidoscope itself. Thus terms are defined in terms which themselves require definition, leading to a yet further round of definitions. The total structure soon passes beyond our conceptual grasp or con-

---

[6] In psychoanalysis, of course, this process is a desirable one, in that destructive and counter-productive contents, once made conscious, may be dismantled.

trol. This, then, is philosophy at its worst. Though the disease is, I think, endemic, there is yet in the great philosophers (as most would agree) a thread, a vein even, of real value. There is that which constitutes a genuine contribution to human understanding. In the case of Schiller's aesthetic *Letters* I am taking it upon myself to draw out that which I personally consider to be of real value. I shall ignore here those aspects for which I have no use. In particular I shall avoid the multiplicity of terms used by Schiller, selecting only those which are in fairly clear accord with my own, avoiding those which might lead into – as I see it – idle 'philosophical' speculation. I would, of course, not simply advise the reader to familiarize himself with Schiller's full account, but would hope that my brief excursion into it would cause him very much to want to do so.

In turning now to Schiller's actual hypotheses, we first need to take up my earlier suggestion of his AB profile : that in Schiller we see a man exceptionally endowed in both major personality dimensions. He would be and is, therefore, as his direct accounts of himself show, and as his theorizing further archestructurally reveals, an example like Goethe and Faust of the man in whose breast dwell 'two spirits'. That is, he is modern (western) man writ large.

The position, as I say, was at no time a secret from Schiller himself. As his editors also comment :

> Schiller . . . saw himself as something of a 'hermaphrodite', a cross between a poet and a philosopher . . . he would complain that when he was trying to write poetry the philosopher got in the way, and when he was supposed to be philosophizing the poet caught him unawares . . . Such ambivalence . . . was but one symptom of a dichotomy which tormented him for the greater part of his life and was, if we are to believe his own testimony, never entirely healed at the end of it.
>
> (*Introduction*, op. cit., p.xxix)

That Schiller's theory of aesthetics is, from one point of view, 'merely' an unconscious reflection of the structure of his own personality does not weaken the theory, as at first might seem the case. First, as I have argued elsewhere, *everything* that man (every man) does is always in large degree a reflection of his own inner structures and processes. Moreover, the man of *one* part (allowing that one part to represent his whole) not only lends a one-sidedness to his accounts of the universe (so the scientist, so the romantic poet) but achieves thereby also a spurious coherence, a one-ness with his subject or object that carries the conviction of wholeness or a full truth. This is to say, we are misled into thinking that the scientist, the accountant or the politician is *not* writing about himself, because he has stepped aside from most other aspects and functions of his personality : except for the one part cunningly concealed (also often from himself!) beneath his overcoat. The truly full man reveals his fullness – but this, paradoxically, is then seen to be either at worst a collection of disparate parts; or at best not a natural, as it were from-the-beginning whole, but a union of parts, a having-come-together, a synthesis of oppositions. (The end result is nonetheless a true synthesis for all its origins in separateness.) So, then, in less adequate men we find a one-ness (a one-sidedness) that passes for wholeness. In more gifted men we find perhaps only an *achieved* wholeness – and more often clear signs of fragmentation – both of which nevertheless are a *real* as opposed to a spurious totality.

Schiller's theory – to turn now to the actual theory – considers the basic personality of man to rest upon two drives. These he calls the 'sensual' and the 'rational' (*Thirteenth Letter*). He also refers to them respectively as the 'intuitive' and the 'speculative' understanding, and in terms of 'all-unifying Nature' and 'all-dividing Intellect' (*Sixth Letter*), and so on. I shall confine myself to the use of the first pair.

It will already be perceived how clearly both in particular and in general Schiller's (of course much earlier) terminology and his approach align themselves with my own System B and System A, with the Self and the Ego.[7] I shall let these next two short extracts speak for themselves. The first arises from a discussion of the origins of civilization and the modern State.

> The intuitive and the speculative understanding now withdrew in hostility to take up positions in their respective fields, whose frontiers they now began to guard with jealous mistrust; and with this confining of our activity to a particular sphere we have given ourselves a master within, who not infrequently ends by suppressing the rest of our potentialities. While in the one a riotous imagination ravages the hard-one fruits of the intellect, in another the spirit of abstraction stifles the fire at which the heart should have warmed itself . . .   (*Sixth Letter*)

The second extract is from Schiller's discussion of why the natural sciences make such slow progress. If we include here psychology and sociology we have, I think, a most trenchant comment on what I personally consider to be the failure of those two disciplines, as at present constituted.

> One of the chief reasons why our natural sciences make such slow progress is obviously the universal, and almost uncontroll-

---

[7] Also with the underlying principles of what I, using Jung's formulation, term Synchronicity, and with Causality respectively.

able, propensity to teleological judgements, in which, once they are used constitutively, the determining faculty is substituted for the receptive. However strong and however varied the impact made upon our organs by nature, all her manifold variety is then entirely lost upon us, because we are seeking nothing in her but what we put into her; because, instead of letting her come in upon us, we are thrusting ourselves out upon her with all the impatient anticipations of our reason. If, then, in the course of centuries, it should happen that a man tries to approach her with his sense-organs untroubled, innocent and wide open, and, thanks to this, should chance upon a multitude of phenomena which we, with our tendency to pre-judge the issue, have overlooked, then we are mightily astonished that so many eyes in such broad daylight should have noted nothing. This premature hankering after harmony before we have even got together the individual sounds which are to go to its making, this violent usurping of authority by ratiocination in a field where its right to give orders is by no means unconditional, is the reason why so many thinking minds fail to have any fruitful effect upon the advancement of science; and it would be difficult to say which has done more harm to the progress of knowledge: a sense-faculty unamenable to form, or a reasoning faculty which will not stay for a content.

(Footnote, *Thirteenth Letter*)

We start, then, with two basic drives. Their effects upon each other – while the personality is still divided – are wholly inimical and destructive. We, in modern western society, are very clear of the destructive and diverting effect of feeling and emotion on thought, and we, in a sense rightly, condemn the intrusion of such (subjective) aspects of our characters into the (objective) tasks of the laboratory. This is a complaint registered of course by the Ego. We are far less sensitive to the destructive effect of thought upon 'feeling' (feeling here being understood to include all aspects of the Self's activities and the operation of the principle of Synchronicity). Schiller writes on these points:

> The pernicious effect, both upon thought and action, of an undue surrender to our sensual nature will be evident to all. Not quite so evident, although just as common, and no less important, is the nefarious influence exerted upon our knowledge and upon our conduct by a preponderance of rationality.[8]
>
> (*Thirteenth Letter*)

---

[8] The second complaint here is that of the Self – less frequently heard, and still less acted upon, in a society where the Self is either quite

Having described his two basic drives in some detail and shown their fundamental irreconcilability, Schiller extricates himself from the apparently unresolvable position he has deliberately placed himself in, by the appeal to, and the demonstration of, a third drive – one which is at once both other drives and neither, which sets no limits to either basic drive, yet incorporates and transmutes both. This idea, which will be clearer from the next extracts, closely foreshadows my own notion of System C.

At first sight nothing could seem more diametrically opposed than the tendencies of these two drives, the one pressing for change, the other for changelessness. And yet it is these two drives which, between them, exhaust our concept of humanity and make a third fundamental drive, which might reconcile the two, a completely unthinkable concept. How, then, are we to restore the unity of human nature which sems to be utterly destroyed by this primary and radical opposition?

It is true that their tendencies do indeed conflict with each other – and this is the point to note – not in the same objectives, and things which never make contact can never collide.[9]

(*Thirteenth Letter*)

---

banished, or at best allowed only the minor key. *Therefore*, only a few of our *legends* describe the effect of System A upon System B *in terms sympathetic to B*. The clearest instance is the tale of King Midas – who turned everything he touched, including everything living, into hard, lifeless gold. A further example is perhaps the tale of Medusa, whose head's glance turned men to stone. This legend, however, also does double duty as one of the myriad tales of the effect of the Self upon the Ego. (Medusa is, after all, a woman – *but* her home is said to be in the far *West*, not the East, where the Ego would instinctively place it.) A third instance of the Ego's effect on the Self is in the 'breaking of the spell', once again a destructive effect. But here the attitude of the Ego is again predominant – for, in general, we tend, though not always, to consider the breaking of a spell as a good thing and the casting of a spell as a bad. (Women, with a certain ambivalence of attitude on the part of men, are often described as bewitching and enchanting, and a man in love with one of them is said to be under her spell.) The Self, of course, holds exactly the opposite view of spells.

[9] An extremely valuable point is made here. To give a concrete example, I take Schiller to be saying that science *cannot* invalidate religion, or religion science, because their aims, methods, premises, content, motivations, etc. *never coincide*. That is to say, they *should* never clash. A problem does begin to arise, however, when the scientist takes it upon himself to tell us what religion is about, or the priest to tell us what science is about

Nevertheless, Schiller is not advocating here a 'Caesar and not Caesar'

On the Aesthetic Education of Man 161

That man is consequently, in the fullest sense of the word, a human being, is never brought home to him as long as he satisfies only one of these two drives to the exclusion of the other, or only satisfies them one after the other.   *(Fourteenth Letter)*

I call a man tense when he is under the compulsion of thought, no less than when he is under the compulsion of feeling. Exclusive domination by either of his two basic drives is for him a state of constraint and violence, and freedom lies only in the co-operation of both his natures.   *(Seventeenth Letter)*

Should there, however, be cases in which he were to have this total experience *simultaneously* ... they would awaken in him a new drive which, precisely because the other two drives co-operate within it, would be opposed to each of them considered separately and could justifiably count as a new drive.
(Fourteenth Letter)

As to precisely *how* the two drives unite (or what can be done to ensure that they do), instead of remaining as separated as they so often are, Schiller has little to say. I shall myself be making a very short attempt on this question a little later. Schiller, however, is in no doubt at all that the almost magical shift described can and does occur – his implicit view being (I think quite rightly) that if you can observe an outcome you are justified in postulating a process that yields it, even if you are in no way able either to describe the details of the process or to reproduce it to order. Some of Schiller's descriptions, however – as far as they go – are these.

The [sensual] drive wants to *be* determined, wants to receive its object; the [rational] drive wants ... to determine, wants to bring forth its object. The [third] drive, therefore, will endeavour so to receive as if it had itself brought forth, and so to bring forth as the intuitive sense aspires to receive.   *(Fourteenth Letter)*

(The outcome – the third product of the combining of the two

---

solution. He is not recommending partition or apartheid in that sense. He *does* say later, that one of the tasks of a society is to define the areas of activity of the two drives – whenever it is desirable or necessary that one or other should be used in isolation. But this is in respect of specific, local or circumscribed phenomena or situations. As we shall see, in all major undertakings – the character and conduct of life and society, for example – it is a joint action, a common and simultaneous engagement of *both* drives he feels, which is the solution, and the only solution.

initial components – Schiller refers to principally as beauty, but sometimes also as Reason, and yet again as freedom.)[10]

> Beauty, it was said, unites two conditions which are diametrically opposed and can never become One. It is from this opposition that we have to start; and we must first grasp it and acknowledge it, in all its unmitigated rigour, so that these two conditions are distinguished with utmost precision; otherwise we shall only succeed in confusing but never in uniting them. In the second place it was said, beauty unites these two opposed conditions and thus destroys the opposition. Since, however, both conditions remain everlastingly opposed to each other, there is no way of uniting them except by destroying them.
> (*Eighteenth Letter*)

> Nature always unites, Intellect always divides; but Reason unites once more. (Footnote, *Eighteenth Letter*)

> [Freedom] arises only when man is a complete being, when both his fundamental drives are fully developed; it will, therefore, be lacking as long as he is incomplete, as long as one of the two drives is excluded, and it should be capable of being restored by anything which gives him back his completeness.
> (*Twentieth Letter*)

Turning again momentarily to *how* this 'miraculous' synthesis occurs, or can be made to occur, Schiller would seem to say, at least implicitly, that as long as neither of the two basic drives is neglected, as long as each is continuously fostered and nurtured (but never *indulged*, and doubly never so at the expense of the other), somehow the harmonious linking or synthesizing of the two will, of itself, and by nature, at some stage occur. I would myself generally concur in this view. I would feel that once attention (of the individual and of society as a whole) is drawn to the essential duality of our nature, and once the consequences of this are accepted, at every level, in the planning and shaping of society, in particular of our educational system, we *may* thereby have created the necessary background conditions within which the harmonious uniting of the basic drives will spontaneously occur. We are dealing with a natural process which we may not be able to force, but only to encourage or facilitate. Should the desired outcome not result, however, given the conditions I have outlined we (man) are then faced, I think, with a very real crisis.

---

[10] Interestingly, I have myself described the characteristic attribute of System C as 'Reason-ableness' (and *not* Rationality); and also, of course, regard the aesthetic or artistic experience as a major manifestation of the operation of System C.

That the 'aesthetic' unifying experience we are discussing *is* however, a fact, and not some hypothetical or imaginary construct, can be ascertained in the first instance, I suggest, from one's own personal experience of art (music, painting, writing, and so on). I have broached this question of identifying the System C experience in the penultimate chapter of *Total Man*. However, I think we can also readily see the literal evidence of the validity of this general concept in another way – *and see that its conditions are indeed met, both literally and figuratively* – in the following examples.

First, being 'in love' (marriage in *this* sense) is a condition in which two creatures different enough (as I have suggested) to be ranked, under slightly altered circumstances, as two distinct species (if not two distinct kingdoms!) relate to each other in such a way as to (1) form a unit and (2) confer upon each other a qualitatively changed experience of life, quite other than that which they experience living in isolation. Additionally, in the act of mating a male and female (with their entirely differing sexual parts and organs of reproduction) interact physically to produce a child, which is at once both of them and neither of them. Third, when we cross two species or varieties, such as a horse and an ass, we produce a creature – in the case cited, a mule – which is at once both and neither of the parental species. And finally, at the purely conceptual level, we speak regularly of thesis and anti-thesis (against thesis) producing a syn-thesis (with thesis). As it happens the word 'with' itself is one of those today rare words, more common in ancient times, which stand for both the thing and its opposite – 'with' meaning both 'for' and 'against' (as in 'with-hold', 'with-stand'). We may set up the various foregoing as triads, thus:

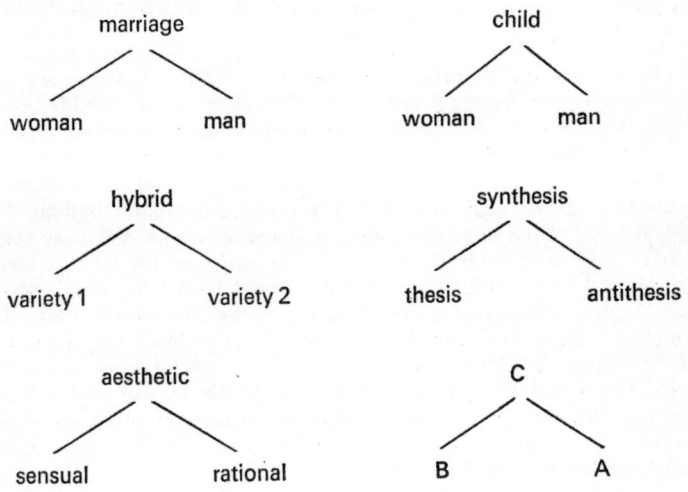

These instances, along with many others I have not included here, are, I suggest, archestructural perceptions and renditions of both the intra- and inter-person psychical conditions we have been discussing.

Before closing this particular exposition, I would like finally to attempt some statement – mainly a pure description – of what seems actually to be going on at the psychological level during the synthesizing experience, and some linking of this with our earlier discussions of consciousness and of symbolic evolution.[11]

(ii)

That specific memories are underpinned by an actual, physiological trace-record somewhere in the brain seems beyond argument. This view would appear to be fully borne out by the fact that destruction or anaesthetization of minute areas of cortex can cause loss of specific memories, recovered when the area is de-anaesthetized, and alternatively by the fact that electrical stimulation of similarly minute areas can *repeatedly* arouse (in the subject's mind) complete, structured memories. Quite what precise physical form this storing takes is, however, a matter for conjecture. One somehow assumes co-operation of clusters of neurones in respect of each memory, rather than any kind of one neurone=one trace arrangement. Contents which have passed through consciousness and been stored by it can become conscious once again. *I* prefer to say 'can be activated by consciousness' – implying, of course, that consciousness has an independent existence of its own quite apart from the existence of any specific or general memory-trace, or group of such traces. The memory-trace certainly continues to exist (possibly indefinitely) even when not in consciousness, and even after not having been brought into consciousness for many years. I use the phrase 'brought into

---

[11] One or two smaller parallelisms between Schiller's formulations and my own are worth bringing out. In the *Twentieth Letter* Schiller comments: 'The sensuous drive comes into operation earlier than the rational . . . and it is this priority of the sensuous drive which provides the clue to the whole history of human freedom.' I myself have also always insisted on the primacy of System B and the Self in my own conceptualizations. Towards the end of the same letter also an image occurs which I have made much of, and not merely figuratively: 'The scales of the balance stand level when they are empty; but they also stand level when they contain equal weights.' Finally, at various points – in the *Seventeenth Letter* for example – Schiller indicates that the aesthetic experience may variously, and for various individuals, be composed of unequal mixtures of the two drives. There will be differing admixtures of the sensual and rational elements. This does not, however, affect or devalue the art-status of the experience in question. Again my own view of art (the System C experience) contains a similar provision.

consciousness' again quite advisedly. Indeed, it is as if consciousness goes to the trace and takes away a copy (as we might take away a photocopy of a book held in a library). For the memory-trace seems not thereby harmed or diminished or (as far as one can see) 'moved', let alone removed, by having been remembered or recalled. We could if we wished remember the same event every day of our lives without using it up or 'wearing it out' – in fact, the memory might well because of this constant rehearsal become even more vivid during the period in question. The memory also seems almost always to be in the same place as last time – at least relative to other memories, and at least in the sense that we have (as a rule) no difficulty in locating it immediately.

Now, whether memories are stored in layers (with the earliest at the bottom) or whether somehow serially (with the first memory as number one) or whether rather in meaningful groupings, or whether in some yet far more complex way, as is no doubt the case, no matter which of these views is most preferable or actually correct, it is clear that no two memories can occupy *precisely* the same space. And even if some are stored close to each other, or even slightly overlap, still most memories will, by definition, be a very long way from most others. In terms of micro-structures only a few molecules 'thick', we must remember, a physical separation of one or two centimetres corresponds to the inter-stellar magnitudes of the astronomical universe. Nonetheless, I can choose any two memories from among my vast store and have them side by side in consciousness any time I will. Does not this fact alone again assure us that the memory or memory-trace is not the same as consciousness? For these two memories are physically (and chronologically) nowhere near each other. And in any case, as we observed earlier, a memory brought into consciousness does not, as a rule, have the effect of disturbing or disrupting the basic physiological trace, unless, perhaps, we wish it.[12]

---

[12] I am, of course, deliberately simplifying the discussion here. For an established memory trace *can* nevertheless be readily changed by the action of consciousness, at least under some conditions. For example, if I have wrongly recorded the name of a friend's wife, as soon as I am told the real name, I am likely to replace the wrong name by it in all relevant situations. I may, of course, then still remember that I once misremembered her, and even remember what the mistaken name was. (But these, I think, are new memories in their own right.) But certainly my already existing memories of the woman have now been edited and the correct name substituted for the wrong one. When I recall or refer to *already established associations*, they now have the right (new) name attached to them.

There is finally also some suggestion that physiological memory traces undergo some kinds of spontaneous change, as well as perhaps straight-

That trace remains not only *in situ*, but intact, again once more suggesting that it and consciousness are not one and the same.

Let us take an example in detail, for this will help prepare us for the rather more ambitious suggestions at the end of this chapter. Let us say I have an idea in my head for a play about a man who kills his wife, but tries to carry on publicly as if she were still alive and living in his house. I have an idea for another play also, invented on a different occasion, about a woman believed dead who realizes that this is her chance to disappear and begin a completely new life. At some point I decide to amalgamate the two ideas into one play – the husband, as before, believes he has killed his wife and carries on as if she were still alive; but she is not dead – and having understood what has happened, takes on a new identity and begins a new life in the same town, where she can watch her husband.

At this point in time I still have, in my head, a perfect record of the first play and a perfect record of the second play, but I also now have the record of the details of the third play, which is, substantially, the other two put together. I can, then, in my head, apparently have my cake and eat it. Perhaps much more importantly still, all these events have taken place only in my head and nowhere else. They have at no time existed outside of it. It is certainly true that they *now* exist in the objective sense as memory traces. But the memory trace exists *after* the event, not before. What kind of existence, may one ask, do ideas in the head have *before* they become memory-traces and before being made public? (And why, may one ask, is the academic and experimental psychologist interested *only* in such historical aspects and not in the continuous, on-going miracle of the production of new ideas – and in *its* mechanics, parameters and significance?) If I make a table, afterwards there is a room full of wood chips, remainders and tools. Where are the remainders and rubbish after I have made a play? Some exist indeed as superfluous memory traces. Yet a good deal of the original material seems to have somehow evaporated. It is somewhat like someone playing for us a piano piece. At the end of the playing I have a memory of the performance. But the sounds, the actual notes have died away and vanished in the air. Yet how often do we find mention of the 'sounds' in the mind – i.e. of ideas, thoughts, the living consciousness – in the current textbooks of psychology?

When I put ideas together in my mind, the process is rather as if I had, say, three boxes – one containing red squares, one containing blue triangles and one containing green circles. I take these from

---

forward decay, without our being aware of it. Thus we may swear that our wife was wearing a red jacket at the time of our first long-ago meeting, while *she* firmly claims to have been wearing a green dress.

their boxes, or some of them only, and have them in front of me. Miraculously, of course, the three boxes still all contain their full complement of blocks.[13] With the blocks that I have removed, but not removed, from their boxes, I now make patterns – patterns that never before existed. Some I keep – that is, remember. That is, I lay down memory-traces of them. In so doing I create new permanent boxes with new permanent bricks in them (this time arranged in fixed patterns) available for future use. But of course I still have all the blocks I started with in front of me.

Yet still this is an insufficient analogy. For sometimes what I make with my materials is really new. It scarcely resembles, if at all, what I took from the boxes. It has changed, it seems, its very nature.

If this now sounds a little too miraculous, let us take a simple, everyday parallel from the laboratory. When I burn hydrogen in oxygen, I find I have neither of these gases, but instead water.[14] So it can be with mental contents (except that, as ever, at the end I *still* have the hydrogen and oxygen I started out with!). All the great insights (mental) and inventions (physical) man produces are of this kind. Something new *does* then exist under the sun. One actual example – though there are many kinds – is as follows. A man sees that a piece of wood resting on an edge of rock pivots easily. He, by pure chance perhaps, 'discovers' that pressure on one end of the bough causes the further end to rise up. But then one man, one day, *realizes* that he has a means of telling which of two objects is truly heavier. For placed at either end of the balanced bough, the heavier one of two objects lifts the other. He has invented a weighing machine, a balance. As a balance, it is now something quite other than a piece of wood and a piece of rock.

Perhaps with this example we are in a position both to put forward, and to see as less far-fetched and 'miraculous' than we otherwise might have done without preparations, the suggestion of Schiller (and of myself some 170 years later) that the two basic drives, the two whole *personae* of Self and Ego – or at least significant portions of these – *can* come together in consciousness and can there by 'joining' form a new mental content, a state of consciousness, that is, offer an experience, which is quite different from either, but is nevertheless made up of both.

One final step. This change, this joining – the aesthetic experience – takes place in consciousness. It is *not* itself consciousness, though it is seen, experienced or apprehended by consciousness. It is

---

[13] The idea, perhaps, which is expressed symbolically in fairy-stories of the purse that is never empty.

[14] Is this not a further example of the mating of two which produces a different third?

certainly very close to consciousness – in many senses of the word close. For it is, so to speak, happily received by consciousness. It somehow has the effect also of causing consciousness to deepen and widen – indeed, till that consciousness can seem to fill the whole universe. The aesthetic experience will also tend, in being apprehended, to facilitate new aesthetic activity or production in the consciousness of the receiver. (So Goethe inspired Gounod, Berlioz, and a host of writers. So Schiller inspired Verdi, Beethoven. So Shakespeare inspires artists of every medium in every age). Finally, when free of the duties and cares of life – and also when we *would* be free of them, or when searching for a meaning to our existence, or to crown some great moment in man's affairs – it is to the aesthetic experience we turn : to music, to the theatre, to literature.

What I wish to suggest herewith is that the aesthetic transformation perhaps occurs *in consciousness and nowhere else* : that is, that it has its being only in consciousness. It does not come out of one of our boxes – nor, really, can *it* as an experience ever be put back in a box : only some *memory* of it can be stored. In a word, I am suggesting that the experience has *no physical or physiological basis* in the narrow and literal sense of these terms. A little like the sound of the piano, it stays only briefly as it comes to birth, and is then gone.

If this *were* the case – that the System C or aesthetic experience exists only in consciousness – then perhaps we see here the first true offspring of symbolic evolution : the first 'creature' or being to have been bred from the interaction of psychological contents alone – and found nowhere else in nature save in the consciousness of man.

9
True and False Alternative Societies

This chapter begins with some unflattering remarks about Liberated Youth. These are followed by some equally unflattering comments on the Establishment, which I hope will both redress the balance and clear me of any charge of bias.

Some few years ago, a friend described to me the behaviour of a group of young people he had observed in the Portobello Road in London. They were extravagantly dressed, both men and women, in long flowing clothes and headgear of attractive colours. By disposition extremely gentle, they moved slowly, gracefully and peaceably among the stalls. But then, disaster. (The words which now follow are not mine, but those of my friend.) One of the girls became separated from the herd. She gazed about her in helpless agitation, her gentle eyes filled with anxiety. In a street not noted for its conventional dress, she nevertheless stood out. Her 'cover' had vanished. She was both alone and exposed. At this point, my friend said, he expected her to throw up her head and give forth a long, helpless 'mooooooo!'

This somewhat unfriendly account nevertheless brings out many of the features of the hippy, revolutionist, drop-out (and incidentally socialist, trades-unionist) spectrum. The belongingness and the need to belong, the fear of both isolation and the individual response, the wish never to be so far ahead or so far behind that real separation is a possibility. There are other items here too – a certain 'herbivorous' quality, for example – but the instance is too narrow a plank on which to build such a detailed structure of comment. We shall, though, be considering other evidence in due course. For the moment, let me merely assuage those who would urgently point out some of the enormous differences between some of the sub-groups I have mentioned – hippies and trades-unionists, for example. I do not seek to deny the many differences. But these are, for me, always over-ridden by the consistencies and similarities which underlie all forms of what all these behaviours are – System B and Self activities.

Now to the promised slur on the Establishment. Are not the wearing of a wig and robes by judges and of uniforms by the military among the most utterly mindless and ludicrous practices of our time? It is almost past belief that grown men – not unintelligent men in many cases – can go through the mummery of putting on a piece of false hair when engaged in the very serious business indeed of attempting to judge the actions, and determine the fate, of a fellow human being. At a moment when this man should be closest to the humanity of the man or woman on trial, he, the judge, *renounces* that humanity, enthrones himself upon a raised dais to show that he is more important than the accused, clothes himself in fancy dress that you may not recognize him as a fellow human being – and possibly detect that what he sits upon is an arse.

Were rebel dissidents really interested in destroying the Establishment (they are not, despite appearances, as I shall hope to show), they would all forthwith begin wearing judge's robes and wigs and no other form of clothing, until such time as that practice was abandoned by the courts. The dissidents *might* do this for a while if it were declared a fashion (as they have done with military uniform) but first they would stylize it in some way – transform it into 'gear' – which would already destroy the whole point. Nor would they continue to wear it once it ceased to be the latest craze, even if their objective had not yet been achieved. We shall be coming to the central role of fashion and vagary in the group and Self society in much more detail later in this chapter.

Similarly, for his part the judge could destroy much of the *psychological* force of a dissident's protest if he would sit in judgement on him wearing a striped kaftan, dragging on a mild joint and holding a flower in one hand. There is no valid reason why he should not do this – for how is, or could, the operation and logic of justice be in any way affected by what people happen to be wearing?

Yet the judge will not do this. For he, like the dissident, cannot expose the posturing of the other side *without first abandoning his own postures.* Neither of the two can (or will) step outside his role and his group identity. Were either side willing (able) to abandon its postures, victory for it would be assured. But it would achieve its aims by the sacrifice of its own existence. Here is the crux of the matter: for their existence, not their aims, is really what counts with both traditional and revolutionary.

What of the soldier? Another grown man playing at dressing up! But not *playing.* No indeed. As serious as any small boy involved in cops and robbers, he will not only probably strike you should you address him as the boy he is, he may even kill you! Though not, on the other hand, if he is an officer – for this man is not actually unintelligent. The situation is rather that he must, and does, continually rationalize his emotional eleven-year-old personality to himself. Therefore he may calmly (not truly calm, however, merely very controlled) order your arrest, imprisonment, fair trial and execution.

Turning at this point to general theory, we may understand further some of the essential differences between System B and System A communities in the central proposal that all the former are governed by *fashion* and all the latter by *tradition*.

With the term 'fashion' I mean to suggest that this type of society (the System B, Self society) is for ever appearing in new forms, though underneath basically unchanging; while the System A or Ego society, though it changes only very, very gradually – and although its actions are always based almost entirely on precedent – does, nevertheless, in its own slow way, move on and evolve. The System B society, we can say, brings old wine in new bottles; while the System A society brings new wine in old bottles. We see in this summarization once again the reversal of polar opposites that we must now not merely expect, but demand, in any descriptions of the qualities of Self and Ego.

What precise examples do we find of fashion (as defined, the ever new outer form) in System B activities? Consider, for instance, the many varieties even of orthodox religion in Great Britain alone that come readily to mind : Catholic, Anglican, Presbyterian, Methodist, Quaker, Baptist, Adventist, Christian Scientist, Unitarian, and so on. Consider, too, the wide variety of Left-wing and people's movements that exist in the world, and contrast this with the relative similarity and uniformity of Right-wing governments in South Africa, Spain, Australia or America.

Naturally we can also consider here *costume* fashion itself – changing at least once every year. Note, too, that this is very much a phenomenon that concerns *women* – and that styles of male dress are traditionally more conservative. (Observe how readily apposite here are these two terms – 'traditional' and 'conservative' – from the general political arena.) That men are currently more fashion-conscious in the West than they have been for some time is not of particular significance – an acceptable swing of the pendulum, or movement within a range of variability. A clear *relative* difference between the majority of women and men in this context, of course, in any case persists.

What, in contrast, do we find in System A activities? Why, among the customs of the aristocracy, for instance – social, political or whatever – a very considerable leaning on tradition, on the practices of the past? As we note, the very title of Right-wing movement is

Conservative. In the paradoxically inclined phrase 'Conservative movement' we have an unconscious description of the nature of System A – i.e. 'very gradual change'. One need only consider how long innovation by government takes (and compare *that* with revolution). A three-year Royal Commission, a four-year committee, established *simply to consider the position*. Meanwhile the situation which is being investigated persists.

What of British law? This, too, is based to a large extent on precedent. 'What did they do last time?' is the question asked. Changes in the law are made only after long heart-searching; and the necessary new legislation may then itself take as much as years to pass through Parliament.

What also of the academic study and the sciences? Here let us not allow the fact that, in science at least, the collected and stored knowledge happens to be actually useful (its existence even a necessary prerequisite for the discovery of further facts) to prevent us identifying the basic System A credo : adopt new positions only in the light of the old. The compulsive act of taking the past along with one – which up to a point, as I say, happens to be desirable and even essential in the physical sciences – is fully exposed in other disciplines : in the study of Latin in the English school system, for example. Until very recently a qualification in Latin was an essential requirement for admission to an arts degree course. What logic decreed that a historian, a geographer, and so on, *needed* to know Latin, any more than any other language? This is far from an isolated example of the tendency I am discussing. I recall that as Modern Language undergraduates at London University not so long ago we were so occupied with the literature of the past that we never read any books by living German or French authors. The end of the nineteenth century was as far as we ever reached, when merit apparently ended.

It is worth our while to undertake a slight detour into the origins of these generalized relationships of the Self and the Ego to the world of events that we have been discussing. I am afraid that in so doing I shall be adding to my list of enemies among women, but that cannot be helped. For I believe that the nature of woman (and, therefore, aspects of the nature of the Self, since I hold the attributes of the Self – as of the Ego – to be sex-linked) has throughout almost the whole of our recent evolution, by which I probably mean a period of some twenty million years, been shaped almost solely by her relationship to the male; whereas the nature of the male, that is, of the Ego, has been largely shaped by the hunt, the battle and interaction with other males. Parts of this process are, of course, still more ancient, reaching back to the emergence of the mammals and earlier.

To develop this proposition we must first consider the process of domestication. Domestication marks a very significant change in the 'style' of evolution, which had proceeded unchanged during the many billions of years which elapsed since life first emerged on the planet. The importance of the domestication of animals was emphasized by no less a person than Darwin himself, and indeed the opening chapter of the *Origin of Species* is entitled 'Variation under Domestication'. Man himself, of course, is an animal under domestication.

In a state of nature, the individual organism – be this a mouse, a fox, an elephant, a tree – is pitted in its struggle to survive against the totality of the rest of nature. Under domestication, however, the organism is removed from what we might call the evolutionary front line, and placed in a somewhat protected position. So man may sit guard over his flock of sheep at night, set a dog for this purpose, or build a fence of thorn. Or he plants his seedlings in the best positions, removing at least grosser competition by destroying all competing plants and weeds. In a time of drought he may water his plants, in frost cover them with dried grass.

Darwin, I think over-subtly, attributes the undoubted increase in variability under domestication to the (unspecified) direct influence of domestication on the sex-organs of the animal or plant. The situation seems to me a far simpler one. Many plants and animals which would never survive under natural conditions (or would even be born at all) can grow into adulthood. An individual plant might, say, be very susceptible to drought. A week without rain would kill this particular individual and with that of course its genetic legacy. But under domestication it is, perhaps, after four days routinely watered by the farmer. Thus its mortal weakness is never brought into play. The individual plant we are discussing is moreover likely (under the laws of correlation of growth) also to possess further abnormal attributes in some way connected with its weakness (say, particularly small leaves or, less obviously, perhaps a paler or a darker colour of flower than is usual). These characteristics *would never be found in nature*, because the plant in question would have died before showing them and passing them on. But these *are* possibly found in the farmer's crop. Additionally, this plant then survives to produce offspring like itself and becomes – at least potentially – a force in changing the long-term character of the species to which it belongs.

Man, of course, is now (largely) the domesticated organism's environment. *He* acts selectively on the organism, in the same way as raw nature did. A particularly pleasant taste, a pleasing colour, docility, trainability, ample flesh or whatever are variously the characteristics which favour survival. These are reinforced in time by, if you will, 'unnatural selection' on the part of the farmer and husbandman.

I wish to propose that the female human being was domesticated (by the male) to a much greater degree than at first was the male, by himself. To an extent the male in primitive human society also withdraws from nature's forces. But from many points of view, he spends still a good deal of his time 'at the front'; while woman spends most of her time away from it.

As early man's life became more complex, a wholly reasonable division of labour occurred, based on, at the time, wholly logical premises. The male did the hunting (and later the fighting) because he was physically stronger and could run faster, while woman, by reason of her relative physical weakness and slower speed – something of a liability on the hunt (and also, I think, a distraction in sexual terms) – stayed at base, preparing the food, curing the skins, looking after the children. The hunt, and even more so the battle, was a dangerous business – in evolutionary terms a potent and rapid sharpener of desirable attributes by the process of natural selection. Those who were no good at the hunt, or did not enjoy the hunt, were killed more frequently, and this is still more true of the battle. Those who survived were increasingly lovers of the hunt or battle, men always and fiercely proud of their prowess, increasingly unhappy or moody when not hunting or fighting.

Let us, however, return to the position of woman. She does, of course, continue to come into direct contact with some natural hazards. But her main 'environment', I suggest, is her relationship with her husband. Without him survival is increasingly difficult for her – especially when eight months pregnant, for instance. Aside from this general circumstances – aside from needing to persuade her man to stay and look after her and her children – she has also to cope with this increasingly dangerous partner of hers on a day-to-day basis. Each generation sees him stronger, ever more quick to employ that strength in 'relating' to his environment – especially to those around him – without his necessarily meaning to do what in fact, I now suggest, he often did, that is, destroy.

In modern times we still have what is termed the 'battered baby' syndrome. That is, a number of children each year still die at their father's hands after a beating – a beating not necessarily intended to kill, but which nevertheless does so.[1] The automatic right of a

---

[1] As chance would have it my evening paper contained this item:
A farm labourer . . . punched a five-and-a-half-year-old girl to death [the daughter of the woman with whom he lived] in a fit of blind rage, when she kept defying him when he told her to go to bed, a court heard today . . . He said in a statement: 'I know I hit her with my

father to beat his children has indeed never been questioned till the most recent times – and survives too, I suggest, in the teacher's right to cane his pupils.

As for wife-beating, we find in folklore frequent enough reference to the process, one being:

> The woman, the spaniel, the walnut tree,
> the more you beat them, the better they be.

Again until very recent times, the 'ownership' of the wife by the husband has not been questioned. In law, and common practice, from earliest history, the husband has enjoyed the right of total disposal over his wife (and once again, over his children). I suggest we see in this circumstance the tail-end of a yet more dramatic (and I think originally archetypal) situation, in which the husband not too infrequently – both with and without such intention – killed his wife and/or offspring. I believe this situation persisted in society for two quite opposing reasons: one, because of the husband's further readiness to kill anyone who interfered, and, two, his actually essentially *good* relations with other men, and theirs with him.

In the general situation I have described,[2] I suggest that women who could 'deal with' their husbands survived (along with their children) significantly more often than those who did so less well, thus causing this behaviour to become gradually reinforced, that is, to become selectively and survivally favoured. What might 'dealing with' involve?

I think dealing with involved first being sexually attractive to men – so attractive as generally to block or swamp other impulses, including the impulse to strike, or strike and kill. Most external attributes of woman, as it happens, are sexual releasers, in some cases very powerful ones. They have, it seems, been selectively augmented over a very long period by the well-known sub-mechanism of natural selection, sexual selection. It is sexual selection which is responsible for many of the more exaggerated and bizarre features of the anatomy of animals – the male peacock's tail, the stag's antlers, the coloured rear-end of the baboon, and so on.

---

fists a lot . . . I just cannot remember how much . . .' Mr. Harvey said there was partial dislocation of the upper spine and death was due to haemorrhage caused by the rupture of internal organs.
(*Evening Standard*, July 3rd, 1972)

[2] Although I cannot go too far into these matters in the present book, it is actually necessary to make some distinction between the life-style and responses of the early hunters and those of the early gatherers. These groups are approximately the Cro-Magnonoids and the Neanderthaloids respectively.

A glance at the vast range of 'men's magazines' (both 'straight' and pornographic), at advertising, even at much genuine art, can leave little doubt in the mind of the sexual role of human female anatomy in the mind of the male. As it happens, there is also other, more objective evidence. Human buttocks, for example, are notable, unlike those of other primates, for being almost spherical. It would seem that the arousing functions of the pronounced buttocks of the female, with our adoption of an upright stance, have been partly taken over by the spherical breasts – which are of course far larger than their child-rearing functions demand. The notably luxuriant hair of the female *mons veneris*, too, is in strong contrast to the general hairlessness of the female body. This seems to have escaped the otherwise depilatory trend (itself probably sexually produced) for two main reasons : one, because the hair here acts as a diffuser and retainer of sexual scent; but, two, mainly, I suggest, because it functions as a false vagina – the actual vagina of course being well out of sight. The shape and rather precise outline of this clump of hair is in contrast to the ragged distribution of pubic hair on the male stomach.

The Women's Liberation movement has latterly made a public complaint out of woman's long-term private complaint that men habitually, and often only, treat woman as a sexual object. This is certainly true, but the point is that it is nobody's fault – and that men can scarcely help responding to woman in this way anymore than they can help being startled by a loud noise. Nature, without consulting women as to their ideas in the matter, decided to ensure the survival of the female (and therewith, of course, of the whole species) by making her into a super-sexual object. Although at times in this book I have been at pains to underline aspects of our freedom from biology, in sexual and some other matters we are, I am afraid, effectively still nature's prisoners.

Recently Elaine Morgan (*The Descent of Woman*, Souvenir Press, 1972) and others have sought to argue a non-sexual basis for the origins of the female form – notably in connection with the 'aquatic' theory of evolution. This theory maintains that man at one stage in his career lived for a period more or less wholly in water. The theory accounts for our general loss of body-hair and the contrary retention of head hair in terms of this aquatic sojourn. In particular, Elaine Morgan argues that the large breasts and abundant head hair of woman supplied hand-holds for the baby (in danger of) drifting about in the water.

What, then, of the longer eye-lashes of women? Are these also for the baby to hang on to? No, clearly these at least are sexual arousers – as evidence the wearing of dramatically long false eye-lashes by woman as part of make-up.

I do not wish to seem to say that any particular physical attributes of women may not have more than one function, or that the second may not be non-sexual. But that the *major* function of all these attributes is sexual seems amply borne out by the evidence.

There are in any case various internal theoretical difficulties for the aquatic theory as a whole. According to the aquatic view we should expect a positive correlation between reduced body-hair and size of breasts in the female. However, Chinese and Japanese women, while possessing the least body-hair of all human varieties (this is true of the males also), also have the smallest breasts. I shall be discussing my own view of these matters in greater detail in a later book. My suggestion for the smallness of the Asiatic female's breasts is based on that variety's closer relationship to the paleoanthropics who, I suggest, adopted the upright stance later than the neanthropics – and never quite as perfectly. Hence the buttocks continued to play their original role rather more effectively for a rather longer period.

To revert to the female's 'dealing with' the male – this very much involved, I suggest, something else also : again something extremely unflattering to the female. It involved *not having any opinions* of your own, but instead being blank and receptive, so that *his opinions then become your opinions*. Further again – this time a positive attribute – it involved excelling in the arts of communication of all kinds (this both leading to and being evidenced by, I suggest, the female's undisputed greater verbal abilities, discussed in Chapter 4). It involved the satisfaction, and in part the creation, of needs in the male that he is then reluctant to abandon or endanger – among many such items the comfort (both physical and psychological) of a home, the ever-ready ear of a willing listener, the balm for a wounded spirit. Does all this seem to paint too inaccurate a picture of woman, and of male–female relationships, as we see them today?

What, generally, of the male? He has learned, partly through observation and bitter experience, partly by the continuous shapings of natural selection, that in the hunt and the battle *the tried and trusted methods work best*. This is not a complete absolute. There is room, and occasional need, for innovation and the innovator. Such innovation is permanently adopted, however, only after it has been adequately tested, and this practice, too, itself has survival value. What does not confer survival, and what is therefore in time, and in general, bred out, is the impulse to wild and uncontrolled acts and foolhardiness; to *not* looking before leaping; to rushing in where angels fear to tread; to going like a bull at a gate. And there is a development in another area. Those also survive best who relate best to their hunting and battle companions. In these particular furnaces, now, are forged the bonds of comradeship, loyalty, interdependence and understanding *between men and men*. These comments refer of course mainly to the relationships between the men of one tribe or nation. What of relationships between *tribes*? What of hostility and battle? Here, what fairly logically comes to the fore, as elsewhere in nature, is the stylized combat or 'limited conflict' – what still today underlies such concepts as the 'rules of war'. While in man as a species we never see the exquisitely wrought combat rituals (or any other kind of exquisitely wrought ritual) that are found in animal species – for human beings can increasingly think about and govern their actions, a capacity leading them to abandonment of plan or normal practice when extremes or alterations of situation warrant – we do have a kind of approximation to the ethological situation, whereby a species saves itself from too much self-damage through the development of 'protocol'. As we still see in historical times, the issue may be decided by a fight between elected champions instead of a full battle, for instance, or after some fighting surrender may be accepted instead of unconditional massacre. The 'laws of chivalry', the 'rules of war', the 'sphere of operations', or the ritual of the duel, the acknowledgement of refereeship and arbitration, respect for your enemy – all these are evidence in historical times of this half or three-quarter realized 'aim' of the fully ritualized combat.

On the general social level, what are we now left with? We have women who understand men quite well from one point of view – that is, from the point of view of man relating to woman. We have here the achievement of love. Women, however, do not really understand the relationship of man to man. (Nor do women especially well understand other women.) Then on the other hand we have men who understand men very well – as both friends and enemies. Here are the achievements of friendship and honour. Men, however, do not understand women. They have never needed to while their physical strength sufficed to settle any issue without subtlety. Neither do men, in general, understand their children. Conversely, and once more for fairly obvious reasons, women do understand children – but this, I suspect again, specifically concerns children as they relate to adults, and not children as separate human beings in their own right.

The 'fashion' response – i.e. the continuously varied response of the female designed as I suggest in the first instance to divert and otherwise hold the attention of the male – initially a sexual response, becomes generalised into an approach to all life, a general attribute of the Self.[3] The 'traditional' response of the male, initially to the hunt and the battle, similarly becomes the basis for a general ap-

---

[3] I consider the story of Scheherezade, who told her master a series of tales lasting a thousand nights to avoid being put to death, to be a symbolic account of this situation. (It is not the only one, of course.)

proach to life – a general attribute of the Ego. The emancipation (to use Tinbergen's term) or transference of a situation-specific response into a general context is by no means an uncommon event (cf. the chimpanzee's lip-smacking behaviour, and the human smile), and I suggest that such emancipation is a tendency at its most marked in the primates, particularly in man.

One sees here how the consideration of purely evolutionary matters can at least seem to go a good way towards explaining features of twentieth-century life and society, which at first glance would appear very far removed from such primitive considerations. However, my chief purpose here, and the reason for the detour which we have made, has been not so much to bring out the further nature of Self and Ego behaviour, but rather *to emphasize the compulsive and predetermined nature* both of Self and Ego societies.

I should perhaps at this stage put forward the conclusions to which I am heading the argument, before going on to complete this general exposition. My belief, then (and in this, apart from the actual terms I use, I think Schiller would have agreed), is that we have so far in the modern world never seen a society that was not either Ego – or Self-based – principally the former. The so-called Alternative Societies of the modern revolutionaries and dissidents are only the latest, fashionable version of the extremely ancient Self society – of which Christianity was one earlier example and, as I believe, the life-style of Neanderthal before the advent of Cro-Magnon was another.[4] The form (the fashionable expression) of that society changes; but its basic message remains unchanged. I shall be *evaluating* these two alternative societies in conclusion. But let us look at one or two further common identifying features of the Self society – less evident than those of the more uniform Ego society (for the deepest tendency of the Self is always to remain hidden), but nonetheless visible to the searching eye. I spoke of the outset of the 'herbivorous' nature of a Left-of-centre group of youngsters. By this I meant that the links with nature and natural life are never far to seek in the Self community. So, for instance, the common names for marijuana are grass, shit, hash and pot. Hash, of course, is an abbreviation for hashish – but the word happens also to mean a dish of simple food. Pot may be considered to mean cooking pot, or, dare I say, chamber pot. The radicals (this word is derived from the Latin *radix,* meaning a root) often apply the term 'grass-roots' to their basic strength. In the 1972 American electoral campaign some McGovern supporters wore badges which read 'I am a grass root.'

---

[4] See in this connection Ralph Solecki's *Shanidar: The Humanity of Neanderthal Man,* an account of the life of a group of Neanderthaloids living in the Middle East some 45,000 years B.P.

My further guess would be, though I know of the results of no objective survey on this point, that there are more vegetarians among socialists than among conservatives. Certainly the several vegetarian restaurants I have personally frequented have appeared to contain a predominance of anti-Establishment customers.[5] This feature has links with the generalized concern for all aspects of the environment and so-called 'natural' foods – the anti-pollution campaign and general concern for nature that is far more vigorously supported by the radicals, and so on. Finally, an example which I think is all the more impressive by reason of its apparent triviality – the question of free school milk for children in Great Britain. Very few moves by the Conservative Government (and Mrs. Thatcher in particular) have produced such an outcry from the Opposition as the decision to end free school milk. It is, I propose, not a question of whether or not the children need the milk – although, of course, they may certainly do. It is rather that one side thinks they *ought* to have it and the other that they *ought not* to. I believe that the Labour movement wants its children to stay children as long as possible (the unconscious reasons I think also behind the ever higher school-leaving age) and wishes to be a good mother – milk of course being the food of mothers; while the Conservative movement wants its children to become men. Note here many contemptuous phrases in common general use such as 'milk-sop', 'milk-and-water', etc. As I suggested earlier, the Self well knows how to go about with children (and in fact prefers them); while the Ego knows only how to go about with men.

The Self is never far from emotion and hysteria in its various forms (be that merely ungoverned behaviour, or the hysterical paralysis of the neurotic). More kindly, we can say that the Self is often too human – while the Ego is not human enough. As far as I can ascertain, for example, the vast majority of incidents of actual physical assault in the Commons over the last few decades have involved a Labour M.P. as agent. In the 'thirties Emmanuel Shinwell crossed the floor of the House and struck a Conservative who had insulted him, damaging his ear-drum. Recently we saw the assault on Reginald Maudling by Bernadette Devlin; and Jeremy Thorpe was, shall we say, rather vigorously jostled by Labour M.P.s after voting with the Government on the Common Market issue. My *Guardian* of June 28th, 1972, conveniently reported that Denis Skinner (Labour) had picked up and thrown a pile of books at Geoffrey Rippon. I can find mention of only one incident

---

[5] Compare here by contrast the (symbolic) role and value of the steak in American society today – and in frontier and 'male' societies generally.

where, conversely, a Conservative M.P. (Leopold Amery) physically attacked (slapped the face of) an Opposition member. I need hardly, I dare say, draw attention in addition to the many uncontrolled hysterical scenes which have accompanied, and regularly always accompany, demonstrations by various factions of the Left in most major European and American cities, demonstrations against apartheid, racial prejudice and authority in general. (The 'demonstrations' by the Nazis before they came to power in Germany were much more controlled, ruthless and effective.)[6]

Let it not be thought that I am casting aspersions. The endemic crime of the Right wing is naturally an opposite one – that of too much and too frequent *dis-passion*, too much objectivity and too great an emotional control.[7] It is this dispassion which permits slums to exist century after century; which allowed and still allows men

---

[6] From I. M. Lewis's book *Ecstatic Religion* (Penguin, Harmondsworth, 1971) I learn that ecstatic states of *religious* possession are far commoner not only among women, but among the downtrodden and dispossessed generally. These, for me, would all be Self- and B-dominants. The frequent inclusion of overtly, let alone covertly or unconsciously, sexual contents in these ecstatic states is also pointed out by the author – a finding, from my own point of view, both expected and welcome.

In this general connection I suggest that the word 'soul' frequently used by American Negroes – as in 'soul-brother', and so on – and the state of being to which it refers, is actually a reference to Self and the qualities of Self. 'Soul' in this sense is hard to define (as one would expect!) but it has among other things something to do with the love and performance of certain kinds of music (soul-music, blues, jazz) or dance, especially of a rhythmic and mournful character – cf. here the meaning of the normal English word 'soulful'. It has to do also with intensities of religious feeling, of longing (for release, for freedom), and of brotherhood and sisterhood with others of like persuasion. Here, then, is the link with socialism. 'Freedom' here means specifically freedom from white oppression and exploitation. But, as always, what is really being asked for is freedom for the Self – freedom of 'soul'.

Since the issue has arisen here, perhaps I should mention that in my view the 'soul' of the Christian religion is usually a reference to *consciousness*. Nevertheless, compare 'What shall it profit a man, if he shall gain the whole [material] world [i.e. have a powerful, successful Ego] and lose his own soul? [That is, his Self]'.

[7] The 'loss of control' response is, in general, far rarer in Ego groups. As I suggested in the Introduction to this book, it is a frenzy rather than a hysteria. Not only is there less loss of control in the former, but additionally it can be observed still to be very goal-directed. B2-dominants, incidentally, would be far more likely to react with a physical blow than would A-dominants (e.g. Conservative Members of Parliament). The latter would prefer a cutting (N.B.) *remark*.

and women to fall ill in our factories of the most appalling industrial diseases; which sends and has sent thousands and hundreds of thousands of men to their certain, pointless deaths in our silly, pointless battles.

I have tended throughout this chapter to refer to the *un*desirable or counter-productive aspects of Self and Ego societies. There *are*, however, desirable qualities in each. But contrary to received opinion and self-evaluation (particularly on the part of socialism) neither has the monopoly of good qualities. The gentleness of some of the Left-wing (or B) groups is to be approved; but against this must be set the atrocities of many others – of the Christian church itself, of the Russian Communists, of the Irish Republicans. The Self contains as much hate as it contains love. And, for its part, the Ego as much fear as courage. It is the hatred in the one and the fear in the other which lead to that for which we (the human community) have no use.[8] The trump card of the Self is Love; that of the Ego, Chivalry and Honour. While these last-mentioned are not currently held in too high esteem (owing to the generally prevailing socialist climate), they offer nevertheless *as* feasible a basis for an acceptable society as does love.[9]

Though both kinds of society have therefore potential desirable aspects, it is nonetheless impossible for me to describe *either* as in any way right or wrong in any absolute sense. This has to do with my view of morality as necessarily involving freedom and, in particular, freedom of choice. The Self society and the Ego society are *both* in their own different ways predetermined, prejudiced and compulsive.

When we (as consciousness) inhabit the Self or the Ego one of the major drawbacks we encounter and suffer is that we have little

---

[8] Thus the Nazi (System A) fears the Jew (System B) before he hates him; while the rebel (System B) hates the tyrant (System A) before he fears him.

[9] These particular alternatives are represented by *able* progressive education and *good* traditional education – not, I hasten to add, by the weak version of the former or the extreme variety of the latter. Some individuals will always prefer – that is, benefit more from – the one to the other. We do not want therefore either type of education alone: neither is absolutely best, neither is suitable for all. Most studies of the matter seem agreed that girls in general, and some boys, benefit best from progressive education; while most boys, and some girls, benefit most from a traditional regime – see, for example, the findings of a study in which I myself was involved: *Four Years On: A Follow-Up Study of Children Formerly Attending a Progressive and a Traditional Junior School* (Longmans, London, 1966). Further variables, however, include the precise subject matter involved, and so on.

choice but to accept both the desirable and the undesirable attributes of the persona in question, as a kind of package deal. This is one, and certainly a major, drawback. A further equally decisive drawback is that we find ourselves unremittingly seeking to destroy the opposed persona – the Ego if we are Self, and the Self if we are Ego. This of course is a doomed endeavour. No matter what problems the mechanics of that attempted destruction involve in the world of external reality the really unsurmountable obstacle is that we ourselves are internally each and always both Self and Ego. We can *never* escape the other partner of our psyche. The polarity, the division exists as a constructional permanency within all of us. For this reason, certainly, it will always, too, in some sense exist in the world around us. For the step we cannot help taking (the naturally wholly abortive step) in our attempt to deal with this internal situation is to project or discover the undesired half into our surroundings. In practice this leads to the wholly paradoxical position that *we must have the enemy.* In seeking to destroy him, we must *also* preserve him – and in seeking to destroy him we *do* preserve him. For we, in turn, present to him just the stimulus or arrangement he requires to go on being (attempting to be) his own particular half-person and to deal with his own internal problem. This was what I meant earlier when I claimed that the dissident does not *really* want to destroy his opponent. Actually, he wants to destroy him *continually* without destroying him *finally.*

The South Africans and the Russians, while they remain unintegrated individuals will never be able to dismantle their prison camps.[10]

As a final piece of evidence, I would like to illustrate the basically polarized attitude-structure of modern (in general A-dominant) society to aspects of 'rightness' and 'leftness', in which are included also the *political* elements of these positions. As a society – that is, as a species – we are, when all is said and done, A-dominants. There are no *true* B-dominants left in existence (unless, perhaps, in a certain sense, women and children are these). There are, of course, large numbers of people who are B-dominants relative to other individuals. I have outlined the relativist nature of Self–Ego polarity

---

[10] I would like here to draw attention to the psychology of those fine individuals in Russia, mainly writers, who have spoken and continue to speak against the regime. Theirs is, of course, an overground, not an underground, protest – an Ego-protest – as we would expect in an otherwise Self society. It is conducted by intellectuals; it is conducted individually; it is conducted in public. In contrast to, say, Angela Davis, Solzhenitsyn has never attempted to hide either his person or his activities from the attention and reach of the authorities.

in the Introduction to this book. Any true B-dominant, however, would see us *all* as A-dominants – that is, principally governed by our Egos. This is fully borne out by the positiveness of our 'right' associations (etymological and otherwise) and the negativeness of our 'left' associations – as we see on pages 184 and 185. A major point I would like to emphasize is that in a true Self society it would conversely be the 'right' which carried the negative associations and the sketchier profile.

I think that were he here today Schiller would very much agree on the point that the only real, viable Alternative Society is, or would be, one which was simultaneously *both* and *neither* of the two varieties we have been discussing. It would be a society in which, because we were free of the *compulsion* exercised by the Self in isolation, or the Ego in isolation, we could select attributes of either at will, and as best suited any particular need of any parti-

Attitude-Structure of Western Society in Respect of 'Leftness'

(socialist) Left
- weak, clumsy, worthless, womanish
  ↓
  unmanning
  castrating
- base, bass (see Chapter 1)
- 'left handed compliment';
  'two left feet', etc.
- underhand, sinister
- mancino (Ital.) = dubious, dishonest
  gauche (Fr.) = awkward, clumsy
  etc.
- cack-handed = literally 'shit-handed.'

Additional attributes obtained from reversing qualities of the right:
bowed, bent, yielding, wrong, emotional, unholy

NOTE: in general European languages derive the word for left from different roots (c.f. English left, German *links*, French *gauche*, Italian *mancino*, Latin *sinister*, etc.); generally the word for right is derived from the one common root (c.f. English right, French *droit*, German *recht*, etc.), at least within broad groups.

cular moment, while wholly dispensing with the undesirable or counter-productive behaviours and attitudes of each persona.

What form, one may ask, would this society take? It is impossible to answer that question, for the nature and further growth of that society would not really be predetermined. We can probably only describe it negatively – that is, in terms of what would *not* be there. For example, I suggest that political activity as such would entirely disappear. Politics can exist only while internal polarity exists, or rather *persists*, in each of us individually. My feeling also is that commercialized fashion as such – particularly in clothing – would disappear largely in favour of the individual's choosing of what suited *him* (or her) best, in general or at a particular moment. Finally, I would suggest that, importantly (social) role would be entirely replaced by function – that is, only function would remain – and that such aspects of (present) roles which are not functional, like the judge's wig, would vanish.

Attitude-Structure of Western Society in Respect of 'Rightness'

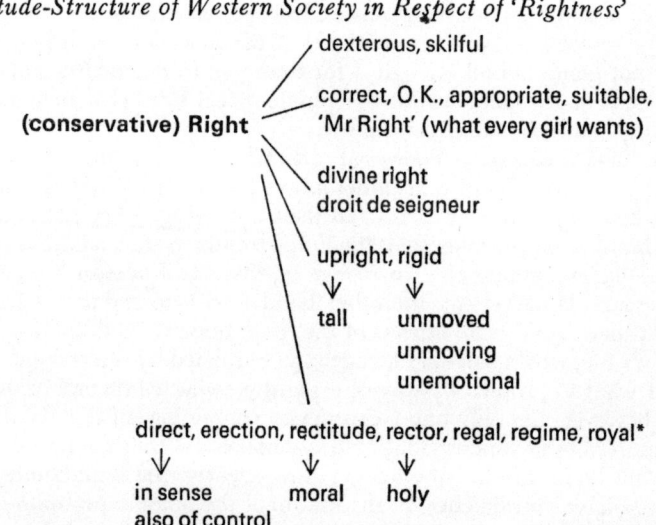

Additional attributes obtained from reversing qualities of the left:
upper hand, good, manly, strong, clean

*All these words are derived from the same etymological root, including the word 'right'.

10
Beyond the Archestructure

I suspect that my view of psychology may seem at times contradictory, in that I on occasion censure its exponents for being too bound to physiology – say, in their denial of consciousness. At other times for not being bound enough – for example, in not taking sufficient account of the influence of the physiological infra-structure on the shape of the conscious personality.

I have a similarly 'equivocal' attitude to philosophy. That is, I feel on the one hand that philosophers take far too little account of the known psychology of man in their theorizing. They fail notably, I think, to appreciate that their apparently transcendant theories are, like everything else we engage in, first and foremost human behaviours. However, on the other hand I wish to say that I believe that there are certain aspects of the (real) universe and the nature of man's life, which can *only* be adequately treated by philosophy.

I want to proceed by way of a slightly awkward metaphor, which is, however, the only one I can devise containing all the elements I require. Let us suppose that man is a house and that the foundations of this house are his physiology. Now, clearly, the foundations of a house have an influence on the nature of the house built upon them. Yet they do not affect all its aspects – the colour of the bricks used, for example, or the detailed styling of the roof. Let us further suppose the house and roof are man's psychology; and finally that the roof of the suggested house is inhabited by birds. The behaviour of these birds is, necessarily, once again to some extent conditioned by the structure and nature of the roof. In many respects, however, they are quite free of it – in the manner in which they fly, for example.

In this metaphor I wish the birds to represent consciousness in general, and conscious ideas in particular. I want further to describe the flight of the birds as 'the play of consciousness' : and propose the 'play of consciousness' as a definition of philosophy.

Philosophy in the sense just defined I believe provides us with information that we can acquire in no other way. In allowing our conscious to wander at will (which is what I consider the philosopher does) I suggest we on the one hand actively discover, on the other hand are somehow passively receptive to, aspects of the universe accessible to us by no other method.

The position I have built up in the preceding paragraphs is offered not by way of validation (for that it cannot be) but at least by way of introduction. For what I now propose is essentially a philosophical, not a psychological, idea. I wish this point to be quite clear at the outset, to forestall its possible use as an objection. I feel even that some will label these following ideas religious or mystical. I do not personally feel them to be such in the conventional sense of those terms. Oddly enough, I think these notions are more likely to prove acceptable to the modern physicist than to modern psychology. These disciplines are currently following quite opposite paths – the psychologist, it seems, into ever greater materialism, the physicist into ever less literally physical realms.

The general concept I propose I term an *epistructure*. The word has perhaps slightly unfortunate associations (unfortunate because none are intended) with such words as 'epilogue'. Nevertheless, the prefix *epi*- is the one I require. It has the meaning of 'over-all' or 'final'. I have headed this chapter 'Beyond the Archestructure'. The epistructure therefore reveals – or, I should say, claims to reveal – still larger forces at work than does the archestructure.

In *Total Man* I drew attention to mankind's tremendous generation of and involvement with opposites and pairs of all kinds in all ages and cultures – mirror-images, twins, identical twins and Siamese twins, mistaken identity, forgeries, shadows, good – evil, left – right, black – white, East – West, art – science, prose – poetry, fact – fiction, and so on idefinitely. Literature, art, philosophy and religion as a whole – these too show man's abiding interest in duality and dualism. I have taken, and still do take, these manifestations to be in the main projections (that is, archestructures) of the major division of man's psyche into Self and Ego, into System B and System A – or, sometimes, as expressions of the further sub-division of B into B1 and B2; or yet again of the polarity which exists within each of these.

Yet 'twoness' or duality cannot perhaps be entirely disposed with as a projection or description of the human psyche. I say 'perhaps' for *some* at least of the items I now mention may in the long run prove to be nothing more than rather persistent archestructures.

That is, they may also be only further projections of human psychology, and not independent facts; persisting, because so deeply imbedded in the very foundations of our vocabulary, and therefore in our attempts to describe the universe, as we think, objectively. I myself though doubt very much that all the items which now follow will be so easily rationalized.

I am thinking here principally of the divisions into inorganic – organic; vegetable – animal; (some aspects of) male – female; the magnetic polarity of north and south; the negative – positive electrical charge; and of matter and anti-matter. Are these pairs 'false' – i.e. purely human – divisions, in actually unbroken continua? Or is there some real polarity of order between them? I am inclined to this latter view. I would myself, for example, propose the difference between inorganic and organic to be the difference between existence and self-experience of existence. Further, if I have correctly understood what I have read on some of these subjects, I believe that many scientists themselves are convinced of the irreducibility (i.e. to one set of referents) of others of these pairs of phenomena.[1]

My question here is, may there not be some overriding aspect or characteristic of the universe which tends to give rise to 'antagonistic' pairs of phenomena?

On general theoretical grounds, I am predisposed to believe that if there were in the universe any overriding force or forces as I shall later describe them, which might lie behind the appearances we have examined, these would be more readily observed by the Self than the Ego. For even if there are any such forces, the Ego will be inclined to say that there are not. The essential ethos of the Ego, as we amply know, is to conquer and to dominate. It is firmly persuaded of its own supra-ordinate place in the scheme of things. On the other hand the Self, by reason of *its* ethos, would be extremely ready to perceive any ultimate influence. Indeed, and unfortunately,

---

[1] Nonetheless, it is well worth bringing out here the extent to which the theorising of the physicist about the universe has revealed, and continues to reveal, the clear influence of the Self – Ego duality of our personality. Thus 'field' explanations of the universe (= Self) have vied and alternated with 'atomic' explanations (= Ego); so too there are wave and particle theories of the nature of light; so too 'steady state' and 'big bang' explanations of the origins of the universe – how closely these last two in particular express the psychological infra-structure. An interesting, and in principle testable, hypothesis arises here. If the pairs of theories I have just listed are (principally) expressions of the duality of the human psyche, then for example male holders of the steady-state view should be on average shorter and stockier than, show less tendency to lose head hair, and so on, than male proponents of the big-bang theory – who should be tall, balding and generally asthenic. Sir Fred Hoyle at any rate (a steady-state proponent) fits the prediction!

the Self is *only too ready* to do so. That is to say, the Self would be and is very prone to *imagine* such influence.

We have a situation where it behoves us to proceed with extreme caution. Quite aside from the question of whether there are or are not forces such as I shall be postulating – when the Ego denies such forces, we cannot trust the denial; and when the Self affirms them, we cannot trust the affirmation.

Notwithstanding, it still remains true that we are more likely to detect the presence of the possible said influences in the products of the Self (and incidentally by ourselves *employing* the powers of Self) than in the products of the Ego, or through the help of the Ego. What now if we examine, say, the myriad forms of religion and superstition which the Self has evolved down the ages. May we, for instance, isolate any universal common elements – distinct from the wide variations in dogma and belief – which religions otherwise reveal in such profusion? I believe that we can. And may such elements be *epistructural* in origin? I believe that they are.

I am not acquainted with all religions. I have examined and compared only those with which I am familiar; and I personally, not some acknowledged authority, have decided that certain terms and concepts in these different religions actually refer to one and the same set of phenomena, despite the differences of vocabulary. (For example, where one religion speaks of 'light' and another of 'spirit' I elect to treat the terms as synonymous; similarly in respect of, say, 'darkness' and 'matter'.) With these provisos, my impression is that religions are consistently concerned with two major, interacting phenomena. This 'twoness' seems to me again unlike the many other 'twonesses' that I have taken to be only projected statements about man's own personality structure. This particular 'twoness' of religion seems to me a genuine, non-subjective phenomenon – i.e. not a man-centred phenomenon, except in the special sense that man himself represents one instance of it.

The central 'story-line' of religion seems to be that what are usually called something like spirit and matter somehow become mingled, especially in the formation of living beings, and notably man. This mixture results in day-to-day life and the universe as we experience it. At death or dissolution, the two combined elements again separate. Linked to the process described is further some idea of progression or development (through time?): perhaps growth in spirituality in one being's lifetime, or an extended growth through many lifetimes, or of a being separated from God or spirit (that is, by imprisonment in the material world) gradually ascending back to God, and so on. The progress described is certainly often cyclic – that is, one which returns in the fullness of time to its point of origin. Nonetheless, progress or movement in some form seems universal in religious conception.

I believe that the aspects of religion I have just outlined refer to a *real* process in a *real* universe – as real, that is, as the formation and death of a star, or as gravity, or as any other scientifically accepted truth. The fact that we find a process described in the (symbolic) language of the Self and not in the scientific language of the Ego *in no way of itself* prejudices the possible validity of the phenomenon.

Turning to my own non-symbolic version of the alleged process, I postulate first a stage in the history of the universe, in which two forces, states or processes are already somewhere and somehow in existence. I say 'somewhere' and 'somehow' because I have no means of describing the nature or the locus of these forces. I think that they in fact lie, at that point, somewhere outside and prior to our own universe. I shall term the two forces 'space' and 'spirit', in inverted commas, to indicate that though other than, they are not wholly unconnected to the normal terms space and spirit and to what is usually implied by these. In a sense space and spirit are the descendants of what I call 'space' and 'spirit', with this proviso : that space is 'space' expressed in terms of 'spirit'; while spirit is 'spirit' expressed in terms of 'space'. Only in these derived or descended forms do we actually apprehend them. I suggest further that the universe as we know it came into existence at the moment the two primal forces began to interact – that, indeed, our universe is the *product* of that interaction.

Perhaps I may remind the reader of the many occasions we have so far found when the interaction of two opposing conditions results in the production of a third condition whose properties could not have been predicted in advance, and which are other than those of the parent conditions. Schiller spoke of the sensual and rational combining to give the aesthetic. In biology, man and woman unite (sexually) and produce the offspring which is both of them, and neither of them. One variety crossed with another variety produces a hybrid. Chemically, oxygen and hydrogen combine to yield a third substance with quite different properties. In physics (while theoretically matter and anti-matter cannot co-exist) the clash of matter and anti-matter *particles* produces a third particle having a stable asymmetric structure.[2] Religion itself, to return to our subject matter, continually speaks of the interaction of spirit and matter combining to produce man – so, in the Judaeo-Christian tradition, God in creating Adam breathes life into clay.

All of the foregoing are either literal examples, or, I suggest, epistructural *perceptions*, of a process which is fundamental, and has

---

[2] Martin Gardner, *The Ambidextrous Universe* (Allen Lane, London, 1967), p. 245.

always been fundamental, to the existence and 'progress' of our universe. I propose that the first result of the first interaction of 'spirit' with 'space' (the entry of 'spirit' *into* 'space'?) was the formation of the simplest physical particles. (Physical matter, then, is 'spirit' in 'space'.)

This physical matter and its descendants make up what man in due course perceives as the external universe. Of course, he is in some sense *himself* a descendant of that first interaction and that first matter. As I conceptualize it, 'spirit' having entered 'space' – having as it were established a bridge-head – does not stop there. It continues – and a choice of phrases is open to us – to move further into, or to move more of itself into, or somehow otherwise to extend its influence on, 'space'. 'Space', however, resists this movement at every stage. 'Space', that is to say, more or less equally determines the way in which 'spirit' progresses. Its resistance perhaps grows in direct proportion – but I think *not in exact proportion* – to the effort made by 'spirit'. For though only slowly, matter develops – that is, becomes continually more complex. As it becomes more complex, so, I suggest, it reflects more of the nature, or at any rate more of the influence, of 'spirit'. The counter-influence of 'space' is, therefore, in some sense gradually lessened. Whether however this counter-influence of 'space' is to be thought of as a quantity which may decrease indefinitely,[3] or whether the decrease is to stop at some point, remains open. In any case the situation is not necessarily to be thought of as a battle (though some kind of struggle or difficulty is obviously involved), where one wins and the other loses. Rather the two forces each continually impose new compromises on each other. And I think – if I may use wholly anthropomorphic terms here – that 'spirit' continually underestimates, continually fails to understand, the nature of 'space', the nature of space, and the nature of itself-in-space, that is, of spirit. There is consequently a great deal of what we must call trial and error, especially, so to speak, at the evolutionary front line. (This notion of a 'God' who is willing but fallible is, I think, a very useful one philosophically with which to approach evolution, and of course human history.) However, despite difficulties and over long periods of time, 'spirit' does seem to achieve a greater occupation of, a greater mastery over, 'space'; and greater understanding of both space and spirit.

I could imagine I have by now entirely exhausted the patience of my more scientific colleagues. I would imagine also that it is equally clear that any attempt to define my terms more precisely, or to apply further my notions to aspects of theoretical physics, would en-

---

[3] Religion implies a yes here – cf. the Christian idea of the Second Coming, the millenium, and so on.

tail not the production of a separate book, but of several – and be in any case beyond my abilities.

As to whether or not I have entirely alienated the scientifically minded individual – I must say this : that whatever its shortcomings, my sketchy proposition does perhaps suggest that the views of science and religion need not necessarily be entirely incompatible. I myself go further and say that they *must* not be. I have always been unable to – and am unable to understand how others can – rest easy with the fact that religious (or Self) accounts of the universe so entirely fail to resemble scientific (or Ego) accounts of, ostensibly, that same universe, from almost every point of view. If the explanations of science are correct, then surely those of religion are the ragings of lunatics. If the religious explanations are correct, then the scientist must be insane. (I am myself in any case entirely of the opinion that any claim to view the universe with any kind of totality *cannot* ignore religious accounts – if only because they are there. They are themselves a phenomenon requiring explanation – and with that will have a place somewhere in the final explanation.)

The sketchy theory I outlined contains for instance an idea of 'movement' that can accommodate, variously, the mystic's notion of progress towards God, the more general concept of (moral?) progress (or purpose), which seems to appeal to many of us very deeply (perhaps epistructurally), and the purely scientific description of change, which attaches no values, posits no goals or direction, but merely tracks the changes.

I am aware that many will find my efforts in these respects farfetched – and even unnecessarily so. This last point in particular I cannot accept. For the far-fetchedness of actuality – that is, of isolated atoms of matter over time having formed themselves not merely into complex organisms, but self-governing organisms endowed with self-awareness –seems to me to outrank any possible degree of far-fetchedness in any theory we might produce to account for the process. My own attempts at explanation at least do not duck, as do those of science and academic psychology, this most central and baffling aspect of organic life – the development of self-awareness. For, compared with *this*, that the interactions of atoms and molecules tend to become more complex over time is a virtual non-event. Yet the lesser aspects are all that science considers.

I myself take awareness (or consciousness) to be not merely far and away the most important event in the universe, but, still worse from the point of view of the sceptic, to be the apparent *purpose* of the whole exercise so far. This, once again, I do not consider to be a religious view, merely a conclusion that my observation warrants.

I close this chapter unoptimistically. The scientific spirit will not so readily be cured of what I and others, consider its myopia in these areas; much less be forced to strike its colours by so tentative a broadside. However, the more broadsides the better. My hope, too, is that better informed individuals than myself may be moved to pursue the kinds of reasoning this chapter has cautiously proposed.

VI

EXPERIMENT AND VALIDATION

Experiment and Validation

Suppose one took the proposition, 'Man is the purpose of the universe.' How would one set about testing it experimentally?

Naturally, the idea of testing this proposition in this way is rather ludicrous. Does that mean however that such a level of enquiry is pointless and valueless? Modern psychology is inclined to think so. Weary (and perhaps not unreasonably so) of centuries of endless and almost entirely unproductive debate on the nature of man, modern psychology took to itself the criterion of testability and the apparatus of the scientific method.

This move produced rapid and valuable advances. So productive in discoveries and insights (compared with the generally unproductive centuries of pure thought on these matters) and so entirely heady were the early days of experiment that something akin to a gold rush was on. At last it seemed we had struck the mother-lode. In a few decades we could hope to know the full and final truth about man. We would predict his every move, cure his every mental ill, satisfy his every dissatisfaction.

There are still today psychologists who believe the foregoing.

Some others of us are unable to escape the feeling that the really essential questions about man and his nature (and about life in general) are not amenable to experimental formulation, nor to experimental testing. With the dust now somewhat settled, we seem to find that experimental psychology is able to deal successfully (but even then, not entirely) with what I have termed the robotic aspects of man's personality – such as conditioning, rote learning, reaction time, and so on. The living, breathing, creating man is somehow lost. In psychology courses all over the world students and teachers sit chewing over crumbs, with the feast itself untouched.

I do not wish to expand here on what, as I have emphasized, I consider the very real achievements of modern experimental psychology. Any good introductory text will describe these to the interested reader. I am concerned, as must be obvious, with the shortcomings of experimental psychology. Let us consider by way of illustration the case of Freud.

Freud is commonly charged by academic psychologists as having *deliberately* expressed his theories in a way which rendered them incapable of testing. There are two points here: (1) that Freud's theories are untestable, and (2) that he deliberately so designed them. The first of the statements is probably true in the broad sense. Such experimental situations as students have succeeded in generating from his ideas bear the unmistakeable stamp of triviality. The grandeur, scope and, above all, the significance of the theory is lost. (This situation derives perhaps largely from the untestable nature of the theory. But in part too, I suggest, not so very much from the lack of imagination of the testers, as from their manifest and latent antagonism to the theory.)

Did Freud, then, deliberately cast his views in untestable form? If one is prepared to say deliberately, but unconsciously, I think the answer is yes. Jung's description of Freud shows a man in an extreme condition of defensiveness, a man in fear for his authority, one who actually faints whenever undesirable truths attempt to force themselves upon him.[1] In my personal opinion, Freud is guilty (though with diminished responsibility) as charged.

I have stated this because with what must by now seem my usual equivocation, I believe that the psychological theorist has an inescapable duty to accommodate the experimentalist whenever he possibly can. In the very nature of events, I think that central, or what one might call integrative, aspects of major theories remain beyond the reach of experimentation and various forms of quantitative evaluation. Yet such theories do (and can certainly be made to) generate what we may call localized spin-off – and these aspects can most definitely be tested in a scientific spirit.[2]

In both this and my previous book I have myself been at considerable pains to state how various implications of my own theory could

---

[1] C. G. Jung, *Memories, Dreams, Reflections* (Collins, London, 1963).

[2] I feel I must add here 'though not in a punitive or antagonistic spirit'. For as I have frequently argued, the *compulsive* scientist (or A-dominant) approaches some phenomena with a wish not to examine them, but to destroy them. Certain aspects of the universe (and of his own personality) are anathema to him. (Jung well describes this position in Chapter 9 of his autobiography, *Memories, Dreams, Reflections*. On a global scale, the new West shows frequently the same response to the old East – cf. the reception given to acupuncture.

be tested. (The more of these that prove negative, the more endangered the central theory – even if *it* cannot be tested directly. Conversely, the more that prove positive, the more strengthened is the theory.) I also assumed, with considerable self-satisfaction, that I would receive credit for this ('Unlike Freud, he ...', and so on!); more to the point, that I would stimulate those with the necessary facilities, which I as a private individual lack, to undertake certain experiments.

However, no mention has in fact been made by reviewers of my experimental suggestions in *Total Man*.[3]

I was both heartened and disheartened by this general reception. Disheartened, because I am myself interested not in winning or losing debates, but in adding to the sum of human knowledge; I had hoped that experimental support for some aspects of my theories would be instrumental in drawing attention to the more central proposal. I was heartened, on the other hand, by the fact that it seemed that experimentalists were apparently unwilling to meet me *even on their own ground*. Might that be because they felt the outcome of the contest was not, after all, a foregone conclusion?

I have therefore not so far been able to discover what experimental psychologists themselves make of Jouvet's finding that cats with the cerebral cortex removed continue to dream, or at least give every sign of so doing. This finding would argue that at least some aspects of dreaming are connected with other parts of the brain and therefore, possibly, with the cerebellum. We also know that female animals with the cerebral cortex removed can still perform the sex act without much loss of fineness of response. *Male* animals so treated cannot perform the sex act at all, even after massive booster doses of hormone. What other part of the brain is then involved in the case of the female?

It was after writing *Total Man* that I came across Breuil and Lautier's comment[4] that in Asiatics the cerebellum is not entirely covered by the cerebrum. Already in that first book, however, I had drawn attention to the experimentally established fact that the cerebellum is on average relatively larger in women than in men, and larger in certain ethnic groups such as Negroes. One looks in vain for mention of any of these items in text-books on psychology – much less for any statement that such gross differences in brain

---

[3] See, for example, H. J. Eysenck, *The Times Higher Educational Supplement*, December 8th, 1972. Only one reviewer observed that what I claimed 'must be refuted not by other words but by evidence that critics may bring to demonstrate the true position that may or may not be different'. (Halla Beloff, *Weekend Scotsman*, December 9th, 1972.)

[4] Op. cit.

structure *might* be accompanied by gross differences in general behaviour.

That women dream more than men and probably sleep more than men is a circumstance one finds grudgingly accepted in an occasional psychology text. I say 'grudgingly' because the matter is then never discussed further. These facts of sleep and dreaming may, as I have suggested, be connected with the relatively larger cerebellum. (As far as I can ascertain possible differences in time spent sleeping and dreaming by different ethnic groups have never been investigated.)

On the question of sex differences in general, I pointed out in the earlier book that anyone seeking information of this kind in mainstream psychology is met with a virtual wall of silence. Paper after paper, book after book contains no breakdowns by sex of the matters treated – even in items such as suicide. Where sex differences are occasionally reported (for example, Woodworth and Schlosberg's comment[5] that the reaction times of females are on average slower than those of males), there is no discussion of the possible ramifications of the finding and least of all for personality theory. No experimental psychologist has so far chosen to rebut this general criticism of mine.

If there is a suspicious silence in current psychological writing on the question of sex differences, there is what can only be called a splendid disregard for all matters connected with left-handedness. One searches index after index in vain. Where any reference is very occasionally made, the subject is accorded a few dismissive lines – almost invariably in the context of physical handicap.

What of observable reality in this case? The observable reality is that, for example, Da Vinci, Goethe, Beethoven, Nietzsche, Michelangelo, and many other outstandingly creative individuals, were left-handed. That, far from being awkward, a sizeable majority of accomplished graphic artists are also sinistral. As it happens, the illustrators of both my previous and this book are left-handers. The latter gentleman tells me that in the last drawing office in which he worked four out of the five graphic artists were left-handed. While the sample is admittedly small, it yields an incidence of 80% left-handedness against a national average of 7%. Moreover, sizeable further numbers of athletes and entertainers again show us that left-handedness is no handicap in any literal sense of that term. Indeed, one's overall impression – it has to remain an impression, for the statistical data is not available – is that the proportion of high achievers who are left-handers in a wide variety of occupations is far in excess of the national incidence of left-handedness of 7%.

---

[5] Op. cit., Chapter 2.

Quite whence, then, the low esteem in which left-handers are traditionally held?

There is a wealth of untapped information on this one subject of left-handedness, of which I have indicated only a few nuggets lying about at the surface; a veritable Atlantis of submerged data. Yet the topic is studiously avoided by modern psychology. Who knows what revision conventional theories of personality and brain function might have to undergo if the subject were to be fully examined? Is this perhaps one of the reasons why it is ignored?

I am sometimes asked, Who might conduct the kinds of research I envisage? And to suggest precisely testable hypotheses. This last I feel I do all the time, but I will be proposing some further instances.

Certainly the task, say, of comparing the relative sizes of cerebella in given human groups is one that could only be undertaken by a medical laboratory. The provision of human specimens is probably costly, and for all I know extremely difficult to organize. Fortunately, my predictions are not only concerned with human beings. I have proposed for example that the cerebellum of the cat will prove relatively larger than that of the dog – for the former we traditionally regard as 'female', and the latter as 'male'. I further propose that the cerebellum of the wolf will be relatively larger than that of the domestic dog. (And do cats with the *cerebellum* removed continue to dream?)

Studies of time spent in dreaming (by either humans or animals) are again properly conductable only in a well-equipped, psychology laboratory. However, in the case of human beings, modified approaches to the topic are possible, through simple questionnaires – a point I will take up in a moment.[6] My predictions here are that short people will in general dream more than tall people; that women will tend to recall more of their dreams and attach greater importance to dreaming generally, than do men – and similarly again for short people v. tall people. I would further predict that individuals with strong religious convictions will dream more (recall dreaming more, and so on) than those with none. A simple interview schedule would ask such questions as these, rating them, say, on a five-point scale of intensity : Do you have the impression you dream a great deal/quite a lot/an average amount/not very much/hardly at all? Do you readily recall your dreams on waking? Do you ever think about your dreams during the day? Do you enjoy your dreams? And so on.

Any national survey of left-handedness would of course cost a

---

[6] Naturally such enquiry would have the shortcomings, the problems of construction and analysis, that bedevil all surveys and questionnaires. These do not render this kind of enquiry completely valueless.

good deal of time, effort and money. Yet any individual school psychologist could obtain from interested headmasters (naturally, in confidentiality) the average height of the left-handers in several schools, and compare it with the average height of all the children in those schools (matching of course for age and so on). For a further prediction of mine is that left-handers are on average shorter than the national norm. Medical practitioners could also be canvassed either by some professional body, or on a smaller scale by simple personal contact (once again in complete confidentiality) to provide details of the height and handedness of their patients.

My theory requires that Socialist Members of Parliament be on average shorter than Conservative M.P.s. I believe this fact is already established – slightly coloured unfortunately by the further fact that there are more Welsh and Jewish Socialists (two rather short ethnic groups). There are also individuals from underprivileged backgrounds in the Socialist ranks, yet another influence on adult height. However, there should prove to be more left-handed Socialist M.P.s – a factor not attributable to the determinants mentioned. The population sample involved here is nevertheless not large, so that I would prefer instead to see a survey of the entire membership of the Labour and Conservative parties. Central Office could in both cases readily mount a postal survey asking for the information quite openly (but, as ever, with complete anonymity for the respondents).

My theory argues that more women than men should be spiritualist mediums (and informed opinion concurs). I suggest again that mediums in general (both male and female) will tend to be shorter than the national average, and that left-handedness will be more prevalent in their ranks than in the population at large.[7] Any of the spiritualist associations could readily conduct these pieces of research. There should be further interlinks, too, between spiritualism, left-handedness and vegetarianism.

I will break off here what will otherwise become a slightly tedious list. I hope, though, to have amply demonstrated, first, that I am far from being an enemy of the experimental approach or the scientific method; and second, that the kinds of experimental validation I envisage are neither all impracticable nor wildly expensive. I am hoping very much that the experimental psychologist will feel unable to side-step this now rather direct challenge.

---

[7] There is however a slight difficulty here. Counter to expectation, fewer woman of normal intelligence are left-handed than men of normal intelligence. (The reverse holds outside the normal range.) I have attempted to account for this unexpected situation at some length in my earlier book (Chapter 10, end). For present purposes we can nevertheless still look for more left-handed female mediums than left-handed female non-mediums.

With that I would like also to close the present book, by adding a word or two about the third volume of the trilogy.

The present book has attempted to open out my theories in what we might call a forward direction. I have tried in general to go a little further into the not readily tangible, and to develop in particular the notion of consciousness as the central actor on what I can conceive of only as a very, very elaborate stage. Consciousness is, in my opinion, the central issue in the study of man; and at the same time the expression of his highest value. The now slightly shop-worn aim of the underground movement – the expansion of consciousness – should, nevertheless, *be* our aim. I include in the expansion of consciousness very much the dismantling of prejudice, which effectively closes off perception from what we might otherwise see – and which others on the other hand manage to see without difficulty. Emphatically, I am not now speaking only of Right-wing prejudice – though that also. I am, for example, more than a little weary of otherwise intelligent Left-wing individuals telling me that height is not genetically transmitted, or that women are not on average shorter than men.

In the third book I shall in the first instance displace the area of focus *back*, to the primates of some twenty million or so years ago. I shall be hoping to show that the foundations of many of our modern problems were effectively laid down at that point. I shall be considering not only prehistoric primates, but as much their present-day descendants. The views I will offer will be seen to have received occasional support from acknowledged authorities – but, like so many of the other items I have so far examined, both then and since been ignored.

The book will also make some attempt, first, to consider the Old Testament as a book of magic, in the sense discussed in Chapter 6. This is no easy task, for we, modern man, stand as far from these magical texts as (to use Ego terms) the cave man stood from the computer, with this difference: the texts lie behind us, not in front of us. Second, I shall attempt to treat the Old Testament as literal history. I shall be proposing, on a psychological level, something akin to the theory of continental drift in geology. That is, I will hope to show that certain large masses of events in early and late pre-history, apparently unconnected and presently thought to be separated by some tens of thousands of years, were actually adjacent; or, more precisely, are the same set of events. As with the theory of continental drift (which proposed that all the present continents of the world were at one time joined together) it is the sheer scope of the events which, I suggest, has prevented till now the recognition of the connections I shall demonstrate. Partly to this end I shall for the first time call the Jewish people to the witness box – a subject deliberately avoided and deferred till this point.

Note on the relationship of libido and aggression

In the footnote at the end of Chapter 1 (p. 58) I proposed a distinction between hysteria and frenzy. This distinction is related to a broader division – that of *libido* and *aggression*.

I consider libido (having adopted Freud's term) as the energy of the Self; aggression as the energy of the Ego. Libido is 'sexual' energy. The term 'sexual' is expressed in quotes to indicate that though the normal meanings of the word sexual are included in the technical term, it does have also less narrowly or obviously sexual aspects. (Here again, I think, I follow Freud.) The term aggression however can be considered to carry much of the same meaning as the everyday term.

As indicated in the first figure below, I believe mysticism to be evolved libido (or 'sexuality'); intellect (or intellectuality) evolved aggression. In this I am making something of the same statement I made in *Total Man*, viz. that emotion is evolved viscera, and thought evolved muscle. We are at the moment considering a rather higher section of the same two continua.

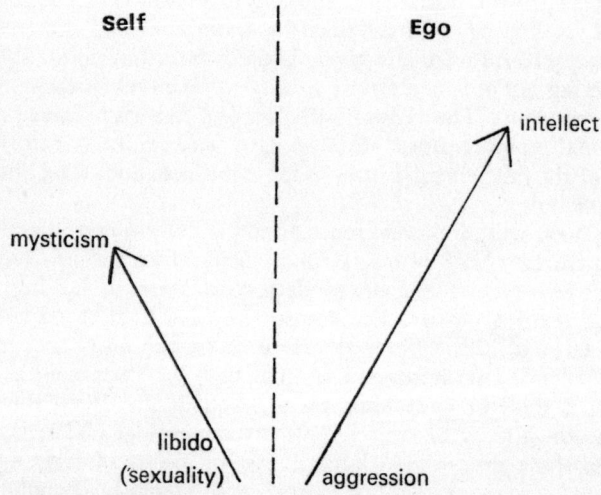

The figure deliberately suggests that in origin (that is, in early evolutionary time) libido and aggression were fairly close; perhaps that they had once even a common origin. (The suggestion somewhat parallels that of Freud's libido instincts and ego instincts, both located in the unconscious – although nonetheless separated there.) In our more primitive moods, the two are still not widely separated. Nevertheless, as the channels or receptors of these two basal energies evolved, their character and their finer expression diverged ever more appreciably. This theoretical position is indicated in the figure by the diverging arrows. Also suggested by the figure is the situation that the evolution of libido is not as fully followed through in *homo sapiens* as is the evolution of aggression. Expressing the position less technically, as a species we are only very poor mystics, though quite well developed intellectually. Neanderthal was a far better mystic – but at the same time a poorer intellectual.

What justification does one have for proposing that mysticism is in fact evolved libido (i.e. 'sexuality') and intellect evolved aggression? The short answer is that in each case the former shows clear vestiges of the latter. Much of the phraseology and preoccupations of mysticism, including religion, are overtly sexual, quite apart from covertly so. Thus nuns are sometimes called the 'brides of Christ', and the mother of Christ is a virgin, who has, however, been impregnated by the Holy Ghost. Again, mysticism is frequently preoccupied with the *control* of sexuality (or with its ritual release under certain conditions, the same coin) which always threatens to become overt or extreme; for example, by ordaining a marriage ceremony, whereby a licence is granted for limited intercourse. The phraseology and preoccupations of cognitive thought and practice on the other hand are for their part often overtly aggressive, let alone covertly. Thus we employ phrases like 'attacking the policy of the opposition' (all that *happens* is that words are spoken) or 'tackling *The Times*' crossword'. The intellect is often concerned with the regulation of aggression, which always threatens to become too overt or uncontrolled. So we have the rules of boxing, the rules of war,

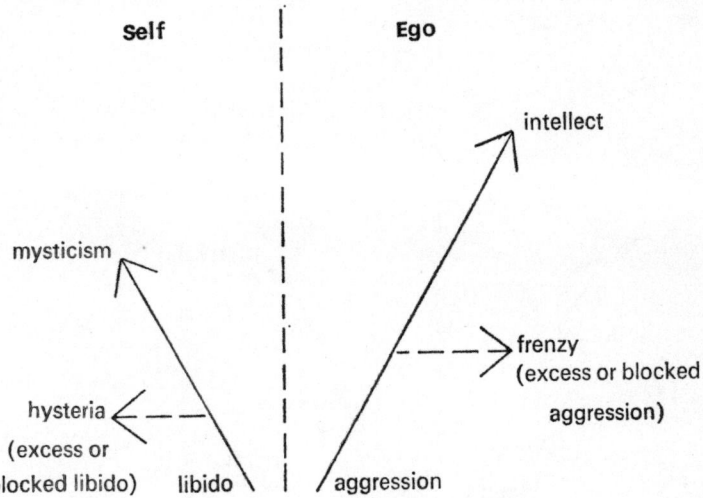

and so on, whereby one is given limited permission to hit or kill another human being.

Any surge of libido is likely to (at least temporarily) destroy and disrupt any mystical or religious framework or activity; any surge of aggression has the same effect on intellectual frameworks. Aside from excesses of energy, it can however also happen that over-strong barriers have been erected (i.e. within the personality itself) against even fairly normal sexual or aggressive responses. (This is effectively Freud's concept of repression.) The arising energy *must* nevertheless be discharged or somehow employed. Yet its proper routes are blocked. The consequences of either too much, or of too drastically controlled, libido and aggression are, I suggest, hysteria and frenzy respectively. The position is shown schematically in the foregoing figure.

I believe it is reasonable to propose that the individual *suffers* hysteria (c.f. hysterical paralysis) but *commits* frenzy – compare such expressions as 'he searched the room frenziedly', 'he drove in a frenzy all night'. Actually normal speech only rather loosely follows my distinction – but I am not relying on common speech here for the evidence, but only for the terminology.

The position I have sketched allows us fairly neatly to 'place' various theorists who have concerned themselves with personality and, I think, to understand the misunderstanding that in general existed among them. Each was actually principally concerned with one part only of the total man. This last diagram ends the present note.

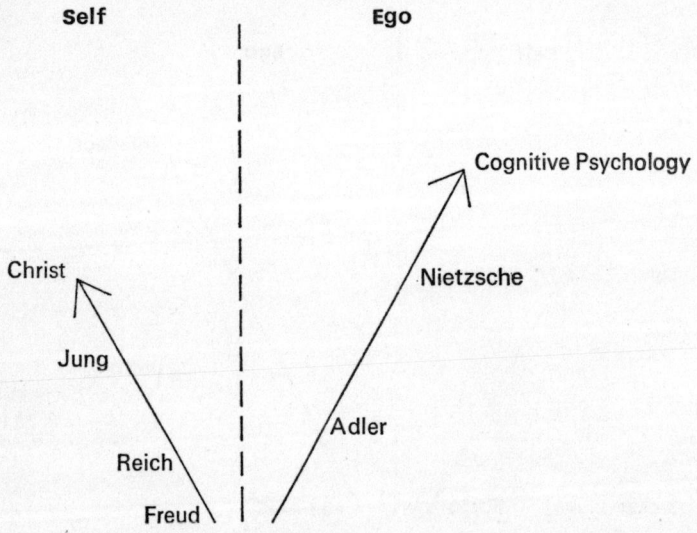

Bibliography

Bettelheim, Bruno, 'Joey: A "Mechanical Boy"', *Scientific American* (March 1959).
Breuil, Henri, and Lautier, Raymond, *The Men of the Old Stone Age* (Harrap, London, 1965).
Darwin, Charles, *The Origin of Species (1859)* (Penguin, Harmondsworth, 1968).
Eibl-Eibesfeldt, Irenäus, *Love and Hate* (Methuen, London, 1971).
Freud, Sigmund, *The Interpretation of Dreams* (Allen & Unwin, London, 1954).
——— *Introductory Lectures on Psycho-Analysis* (Allen & Unwin, London, 1933).
——— *The Psycho-Pathology of Everyday Life*, Complete Works Vol. VI (Hogarth Press, London, 1966).
Gardner, Martin, *The Ambidextrous Universe* (Allen Lane, London, 1967).
Gooch, Stan, and Kellmer Pringle, M. L., *Four Years On: A Follow-Up Study at School Leaving Age of Children Formerly Attending a Progressive and a Traditional Junior School* (Longmans, London, 1966).
——— *Total Man: Notes Towards An Evolutionary Theory of Personality* (Allen Lane, London, 1972).
Goodall, Jane van Lawick-, *In the Shadow of Man* (Collins, London, 1971).
Hall, C. S., and Lindzey, G., *Theories of Personality* (Wiley, London, 1957).
I Ching (The Book of Changes), trans. Richard Wilhelm and Cary F. Baynes (Routledge & Kegan Paul, London, 1968).
Jesperson, Otto, *Language: Its Nature, Development and Origin* (Allen & Unwin, London, 1969).
Jung, C. G., *The Archetypes and the Collective Unconscious*, Collected Works, Vol. 9, Part I (Routledge & Kegan Paul, London, 1969).
——— *Memories, Dreams, Reflections* (Collins, London, 1963).
——— *Synchronicity: An A-Causal Connecting Principle*, Collected Works, Vol. 8 (Routledge & Kegan Paul, London, 1969).

Klein, E., *A Comprehensive Etymological Dictionary of the English Language* (Elsevier, Amsterdam, 1967).
Kretschmer, Ernst, *Physique and Character* (Harcourt, New York, 1925).
Laing, R. D., *The Politics of Experience and the Bird of Paradise* (Penguin, Harmondsworth, 1967).
Leary, Timothy, *The Politics of Ecstasy* (MacGibbon & Kee, London, 1970).
Lewis, I. M., *Ecstatic Religion* (Penguin, Harmondsworth, 1971).
Lorenz, Konrad, *King Solomon's Ring* (Methuen, London, 1961).
Morgan, C. T., and Stellar E., *Physiological Psychology*, 2nd Edition (McGraw Hill, New York, 1950).
Morgan, Elaine, *The Descent of Woman* (Souvenir Press, London, 1972).
Morris, Desmond, *The Naked Ape* (Jonathan Cape, London, 1967).
Reich, Charles, *The Greening of America* (Allen Lane, London, 1971).
Schiller, Friedrich, *On the Aesthetic Education of Man (1793)*, trans. E. M. Wilkinson and L. A. Willoughby (Oxford University Press, London, 1967).
Sheldon, W. H., *Varieties of Human Physique* (Harper, New York, 1940).
——— *Varieties of Human Temperament* (Harper, New York, 1942).
Skeat, W. W., *An Etymological Dictionary of the English Language* (Oxford University Press, London, 1966).
Solecki, Ralph S., *Shanidar: The Humanity of Neanderthal Man* (Allen Lane, London, 1972).
Sparks, John, *Bird Behaviour* (Hamlyn, London, 1969).
Tinbergen, N., *The Study of Instinct* (Oxford University Press, London, 1969).
Woodworth, R. S., and Schlosberg, H., *Experimental Psychology* (Methuen, London, 1955).

Index

ACTOR, 148-9
Adler, A., 206
aesthetics, 153-64, 167-8
aggression, 6, 32, 34-8, 42, 57-8, 73, 87, 89, 120-21, 131, 204-7
Alternative Society, 43, 169-85
America, 25 n, 57, 63, 121, 179-80, 180 n, 181 n
amnesia, 137-8, 146, 148
Anders, G., 130
archestructure, 25-7, 39, 40, 48, 50-59, 60-62, 64, 65-6, 70-71, 78, 90, 115, 118 n, 123, 146, 147, 148-9, 157, 164, 187
 Ego-archestructures, 59, 60-61, 64, 71-9
 Self-archestructures, 65 n, 71, 74
archetype, 25-8, 39, 47-50, 59, 73, 103-4, 108, 110, 122, 145-6, 175
 Ego-archetypes, 59, 60-70, 73, 145
 Self-archetypes, 66, 145
art, 39, 43, 49-50, 59, 67, 125, 129, 162 n, 163, 164 n
artist, 32, 43, 59, 108-9, 153-4, 200
Asiatic, 31, 177, 199
Atlantis, 14, 201

BALANCE, 39-41, 43
becoming, 51, 106-11
Beethoven, L., 153, 154, 168, 200
behaviour, 5, 28, 38-9, 41-2, 48, 50, 52, 58, 59, 60-61, 67, 69 n, 87, 96-104, 113-5, 116-7, 129

being, 34, 37, 135, 137, 148, 149-52
Beloff, H., 199 n
Bettelheim, B., 72
Bible, 24, 30 n, 39-40, 125-6, 130, 203
bird, 10, 11, 109-10, 112-3
body-types, 33, 37, 52-3, 55, 57, 68, 83-90
boss, 70-71
brain, 9 n, 10-14, 28-9, 31 n, 32, 114-5, 146-7, 164, 199, 201
Breuil, H., 31 n, 199
Browning, Elizabeth Barrett, 124-5
Browning, Robert, 107 n, 108
Burt, C., 114
business, 25, 40-41, 66, 69-71

CANING, 175
capitalism, 25, 42, 66, 69-71
cat, 55-6, 57, 58, 67, 199, 200-201
cathedral, 37
causality, 17-19, 97
cerebellum, 10-15, 19-21, 26-8, 31 n, 37, 50-51, 199, 200-201
cerebral cortex, *see* cerebrum
cerebrum, 10-15, 19-21, 26-9, 50-51, 56, 164, 199
chess, 61-2, 94, 114
child, 23, 48-9, 72, 78-9, 87-8, 93-6, 97, 104, 110, 121, 123, 124, 125, 141 n, 145, 151, 163, 170-71, 174-5, 176, 178, 180, 182 n, 183

INDEX

chimpanzee, 64, 109, 110 n, 114, 120 n, 130, 179
Clarke, A. C., 78
clothes, 68-9, 99-100, 148-9, 170-71, 185
cognition, 5-7, 12, 20, 24, 25, 43, 61 n, 67 n, 77, 93-6, 97, 137, 143, 146, 205
collective unconscious, 47, 145
comedy, *see* drama
Communism, 23-4, 41-2
computer, 10, 71-2, 73-9, 117, 138, 145
concentration camp, 57, 183
conditioning, 6, 138-44, 152
consciousness, 7-8, 9 n, 13-17, 19, 22-3, 26, 27 n, 33, 43, 47-50, 72, 74 n, 75-9, 97, 109-10, 115, 117, 126, 135-52, 154, 156, 164-8, 182, 186-7, 192, 203
Conservative Party, 24, 154 n, 171-2, 180-85, 202
creativity, 32, 43, 154, 200
critic, 153
Cro-Magnon man, 29-32, 36, 51 n, 53-4, 55, 84, 108, 115, 175 n, 179
cross, 15, 26
cuckoo, 100-101

Darwin, 27, 56, 66 n, 88, 97, 101, 105, 111, 115, 117, 118, 121, 130, 173
Descartes, 135, 149 n,
devil, 6, 14
Devil, 6, 15, 21 n, 30 n
dinosaur, 101
'divided self', 40
dog, 10, 55-6, 57, 201
domestication, 88 n, 173-4
double-acts, 55
drama, 52-5, 84, 153-4, 168
dreams, 7, 13, 22-5, 37, 39, 47, 56 n, 88, 104, 144-6, 146 n, 150-51, 199, 201
duality, 6, 50-51, 52-9, 60-62, 63, 65, 125, 162, 183, 187-8, 189-93
duel, 63

dwarf, 14, 30

East, 34, 51, 198 n
economics, 69-71
education, 25, 162, 172, 182 n
ego, 6-7, 25, 33 n, 40, 60-61, 66, 87, 204
Ego, 6-8, 14-22, 25-7, 33-4, 38-43, 51-9, 60-63, 65-6, 69-74, 78-9, 89-90, 125-9, 144-5, 146 n, 149-51, 154, 158, 159, 159 n, 167, 171-2, 179-85, 188-90, 203, 204-7
Eliot, George, 35 n
emotion, 5, 34-6, 39, 42, 50, 61 n, 72, 73, 159-61, 180-82, 203-6
epistructure, 187-93
E.S.P., *see* extra-sensory perception
ethology, 47-9, 58, 67, 97-101, 104, 106-8, 109-10, 112-3, 178, 179
etymology, *see* philology
evolution, 5, 6, 12-14, 17, 19-21, 24 n, 25, 27-32, 37-8, 43, 48-50, 51 n, 63, 64, 66 n, 68, 69 n, 72-3, 78, 83, 86, 88, 90, 93, 96-104, 105-8, 111, 114-5, 116-23, 127-8, 130-31, 136, 146-7, 151-2, 172-9, 191
existence, 135, 150
extra-sensory perception, 18, 21, 37-8, 88
Eysenck, H. J., 199 n

fairy-tales, 15, 24, 40, 167 n
fascism, 25, 42, 57, 58 n, 73, 171, 180-85
fashion, 170-71, 178, 179, 185
Faust, 32, 60, 125, 157
female, 6, 7, 24, 27, 33, 35-9, 43, 55-6, 57-9, 65-9, 86-7, 88 n, 98, 99-100, 104, 107-8, 110, 125, 128 n, 154 n, 160 n, 163, 171, 172, 174-9, 180, 181 n, 183, 184-5, 188, 199-200, 202
fetishism, 69, 99
Follower, 6, 21, 65
freedom, 40, 57, 83, 130-31, 150-52, 161-2, 176, 181 n, 182, 184

Index

frenzy, 58 n, 181 n, 204-7
Freud, S., 6, 17, 19, 24, 33 n, 40, 58 n, 64, 121, 152, 156, 198, 204-7
Freudian slip, 13, 22, 23, 51

GAMBLING, 25
genetics, 27, 38, 47, 83, 102, 105, 107, 114, 145, 172, 173
genius, 32
Gestalt psychology, 18
ghost, 6, 14-15, 149 n
Goethe, J. W., 109, 153, 154, 157, 200
Goodall, J. L., 109, 130
Greece, 52-5, 84

HAIR, 68-9, 86, 99-100, 176-7
 body hair, 68, 86, 99, 176-7
 head hair, 69, 86, 100, 176
Hall, C. S., 84
hate, 34-5, 42, 147-8
heredity, 49 n, 67 n, 114
hero, 63, 73, 108, 145
Hippocrates, 84
homo sapiens, 27, 60, 97, 205
homosexuality, 65
Hull, C. L., 141, 143
humour, 39-40
hunting, 25, 62-3, 64-5, 69-70, 172, 174, 175 n, 177-8, 179
hybrid, 32, 86, 90, 163
hypnosis, 23, 88
hysteria, 24, 58 n, 180-81, 204-7
hysterical paralysis, 58 n, 180, 206

I, 147-52
I Ching, 18, 126, 127 n, 129, 130 n
idea, 76, 94, 116-21, 125, 127, 129, 131
ideal, 99, 102-4, 106-10
instinct, 67, 110, 112-3, 120, 138
intellect, 39, 67 n, 126, 129, 156, 204-7
intelligence, 38, 84 n, 95-6, 202 n

JERICHO, 32
Jesperson, O., 94-6, 110, 123 n
Jews, 57-8, 201, 203

'Joey: A "Mechanical Boy"', 72, 73, 79
Jouvet, M., 199
Jung, C. G., 7, 8, 10, 17-18, 24-6, 39, 47, 67, 121, 143, 152, 158 n, 198, 206
justice, 39

KLEIN, E., 57
Kretschmer, E., 84-6
Krishnamurti, 34 n

LAING, R. D., 40, 135, 149 n, 156
language, 93-6, 97, 112-3, 116-7, 121-5, 130-31, 137, 177, 184-5, 187
Lautier, R., 31, 199
Leakey, L. S. B., 28 n
learning, 6, 94-6, 103, 139-44
Leary, T., 43
left, 15-17, 26 n, 38 n, 50, 56, 122-3, 154 n, 171, 179, 184-5
Left wing, 6, 24, 26, 123, 154 n, 171, 179-80, 182, 184-5, 202
left-handedness, 15-17, 26 n, 38, 50, 56, 57, 122, 154, 184-5, 200-202
legend, 6, 13-14, 26, 30, 40, 47, 56, 90, 108, 125, 146, 160 n
Lenin, 24
Lessing, E., 109
libido, 33 n, 58 n, 87, 204-6
Lindzey, G., 84
literature, 5, 47, 108-9, 168
Lorenz, K., 26, 112
love, 23, 25, 34-6, 40, 42, 57-8, 77, 124, 147-8, 163, 178, 182

MACHINE, 71-9, 94, 96, 149 n
magic, 6, 14-15, 25-6, 29, 114, 126 n, 129, 160 n, 203
male, 6, 8, 24, 27, 33, 35-9, 43, 55-6, 57-9, 65-9, 86-7, 89 n, 98, 99-100, 104, 107-8, 110, 125, 128 n, 145, 160 n, 163, 171, 172, 174-9, 180, 180 n, 184-5, 188, 199-200, 202
male society, 65, 174-9
mammal, 10-13, 20, 58, 128, 172

INDEX

man (gender), *see* male
man (species), 10, 14, 18-22, 27-8, 38, 52-6, 60, 71-9, 83-4, 109, 114-15, 125, 130-31, 135, 151-2, 168, 173-9, 187, 202-3
mandala, 39
manic-depressive, 54, 85, 89-90
Marx, K., 24, 42
masochism, 57, 71, 151
mediation, 137-8, 149-51
meditation, 139-40, 146
medium, 24, 88, 202
memory, 22, 117, 137-46, 164-8
mental illness, 6, 41, 43, 49 n, 54, 58, 58 n, 61 n, 72-4, 85, 89-90, 96, 121
mid-brain, 10 n, 12, 20-21
middle classes, 54, 154
millenium, 24
mind, *see* consciousness
moon, 14, 26, 71, 90
morality, 40, 152, 155, 185, 192
Morgan, C. T., 20, 27 n
Morgan, E., 176-7
Morris, D., 25, 56 n
motor-car, 66-9, 75
music, 56-7, 58, 114-5, 153, 168, 181 n, 184
mutation, 102, 105-6, 115
mysticism, 26, 40, 51, 187, 192, 204-7
myth, *see* legend

NATURAL SELECTION, 55-6, 98-104, 173-9
Neanderthal man, 28-31, 51 n, 53-4, 55, 57, 84, 115, 129, 154 n, 175 n, 179, 205
neanthropic, 29, 31, 51, 52-4, 56, 62, 63
Negro, 31, 56, 57-8, 87 n, 181 n, 199
nervous system, 6, 8-10, 20, 22, 27, 32, 50, 52, 122 n, 136, 138, 146, 150
   autonomic system, 8-9, 68, 152
   central system, 8-9
   para-sympathetic system, 32
   sympathetic system, 32

neurosis, 6, 22-4, 41, 43, 51, 58 n, 90, 120 n, 150
night consciousness, *see* unconscious
number, 93, 126-8, 130
numerology, 126 n, 127 n

OPPOSITES, 6, 7, 15-17, 26 n, 39-41, 65, 66, 94, 123, 163, 171, 181, 187-8, 190
organism, 9, 33, 47, 50, 60, 72, 74, 75-9, 83, 93-4, 96-7, 101-3, 105-7, 110-15, 116-7, 118, 121-2, 125, 127-8, 130-31, 135-6, 145, 151-2, 173-4, 188

PALEOANTHROPIC, 28, 31, 51, 52-4, 56, 64
paranoia, 89
Pavlov, I., 121, 142
penis, 64, 68
Person, 42-3, 90, 125, 149
personality, 5-8, 19, 21, 22, 33, 39, 40 n, 41-3, 51, 56, 60, 65, 71, 83-90, 111, 125, 148-9, 152, 157-9, 186, 200, 206
philology, 54 n, 56-7, 65, 122-3, 127 n, 179, 184-5
philosophy, 18, 106, 155-7, 186-7, 191
physics, 187, 188 n, 190, 191
physiology, 8-16, 22, 26, 27, 32, 48-50, 83, 101, 102, 111, 115, 140, 144, 164-8, 186
pineal gland, 13, 20-21
pleasure principle, 17 n
polarity, 6, 34-5, 51, 171, 183, 185, 187-8
politics, 6, 41-3, 50, 59, 118-9, 154, 179-85
pornography, 99-100, 107-8, 176
possession, 6, 61 n (*see also* trance)
primates, 27, 64, 109-10, 151, 176, 179, 203
proletariat, *see* worker
psychoanalysis, 6, 7, 24, 39, 127, 156 n

psychology, 5-12, 15-19, 25, 27, 32-8, 42-3, 47-51, 54, 57-8, 64, 67 n, 71, 83, 87, 93, 101, 111, 115, 121, 123, 125, 127, 130 n, 135-6, 138, 146, 156, 164-8, 170, 186, 188, 197-202
  experimental, 25, 27, 47-8, 121, 123, 127, 141-3, 166, 197-202
psychopath, 60
psychosis, 6, 41, 43, 61 n, 72-4, 85-6, 89-90
pyramid, 37

REACTION TIME, 66-7, 87, 200
reality principle, 17, 51-2, 64, 73
reason, 43, 73, 162
Reich, C., 25, 43
Reich, W., 206
relativity of personality functions, 41-2
releaser, see sign-releaser
religion, 6, 23-4, 26, 31, 34-5, 40-41, 42, 47, 58, 108, 115, 118-9, 125-6, 131, 160 n, 171, 181 n, 182, 184-5, 187, 189-93, 204-5
  ecstatic, 24, 58, 181 n, 205-6
repression, 206
reptile, 10-14, 14 n, 20, 101, 107 n, 128
response, 26-7, 47-50, 58, 66-7, 74-5, 96-104, 105-8, 111-2, 114-5
Right wing, see Conservative Party, Fascism
robotic functioning, 22, 138-43, 149n, 150, 152, 197
romantic movement, 154, 154 n, 156
Romer, A, S., 28 n
Rommel, E., 35 n
Rorschach test, 51
Russia, 182, 183

SADISM, 57-8
sado-masochism, 58
Satan, see Devil
Schiller, F., 19 n, 153-64, 167-8, 179, 184, 190

schizophrenia, 61 n, 72-3, 85-6, 88, 89-90
Schlosberg, H., 66 n, 142, 200
science, 5, 19, 25, 39, 62 n, 77, 111, 118-9, 121, 123, 126-30, 135-6, 157, 158-9, 160 n, 172, 191-3, 198
scientific method, 25, 123, 129-30, 202
second sight, 22 (see also extra-sensory perception)
self, 6-7
Self, 6-8, 12-22, 25-7, 29-30, 33-4, 38-43, 50, 52-9, 60-66, 70-71, 73-4, 78-9, 89-90, 116, 125-30, 144-5, 146 n, 149-51, 154, 158, 159, 159 n, 164 n, 167, 170-72, 178-85, 188-90, 204-6
sex, 24 n, 27, 32, 40, 58, 58 n, 65, 68, 98, 99-100, 101, 104, 163, 172, 173, 175-7, 178, 181 n, 184-5, 204-7
  differences, 6-8, 24, 27, 35-9, 55-6, 66-7, 86-7, 99-100, 107-8, 124-5, 171-9, 199-200, 202-3
sexual selection, 98-100, 101, 175-7
Shadow, 6, 7
Shakespeare, W., 53-4, 153, 168
Sheldon, W. H., 83-4, 85 n, 86-90
sign-releaser, 26, 27 n, 47-9, 58, 67, 97-101, 103-4 106-8, 112, 175-7
Silas Marner, 35 n
Skeat, W. W., 56-7
Skinner, B. F., 121
sleep, 7, 23, 23 n, 32, 87, 88, 117, 200
sleeping consciousness, see unconscious
sleep-walking, 23
social role, 149, 170-71, 185
socialism, 24, 26 n, 123, 154 n, 155, 169, 171, 179-85, 202
society, 43, 54, 55 n, 69-71, 148-50, 155-6, 159, 160 n, 162, 169-85
Solecki, R. S., 30 n, 179 n
Solzhenitsyn, A., 183 n
soul, 30, 40, 94, 181 n
space, 190-91

INDEX

'space', 190-91
species, 6, 28, 38, 43, 47, 56, 60, 88, 90, 99, 104 n, 106, 114, 117-22, 127-9, 131, 163, 173, 183
speech, 94-6
spirit, 190-91
'spirit', 190-91
spiritualism, 24, 88, 202
Stellar, E., 20, 27 n
Stevenson, R. L., 21
stimulus, 47-50, 66-7, 96-104, 105-8, 111-2, 113-4
subjectivity, 19-20, 43, 96-7, 106-111, 159
sun, 26, 51, 71, 74, 90, 135
survival, 30, 38, 62-3, 66 n, 98-104, 106-7, 118-23, 173-9
symbolic evolution, 104, 130-31, 164, 168
symbols, 6, 13-15, 26, 39, 56, 62 n, 69, 178 n, 180 n
synchronicity, 17-20, 126, 129, 158 n, 159
synthesis, 39-43, 158, 162-4, 167, 190
System A, 8, 22, 24-8, 32-8, 40-43, 50, 61, 73, 86, 88-90, 108 n, 154, 157-8, 171-2, 181 n, 182 n, 183-4, 187, 198 n
System B, 8, 22-8, 32-8, 40-43, 51, 56, 58, 60-62, 64, 86-90, 108 n, 130 n, 154, 157-8, 164 n, 170-71, 181 n, 183-4, 187
System B1, 31-8, 86-90, 130 n, 182, 182 n, 187
System B2, 31-8, 61, 86-90, 130 n, 181 n, 187
System C, 40-43, 125, 154 n, 160, 162 n, 164 n, 168
System X, 37-8

TEMPERAMENT, 84, 87
threes, 40, 84-90, 167 (*see also* triads)

time, 71, 72, 138
Tinbergen, N., 26, 78, 96-7, 106, 179
Total Man, 5, 19, 23, 26, 27, 38, 41, 59, 61 n, 88, 123 n, 128 n, 163, 187, 199, 204
tradition, 171-2, 177, 178-9, 182 n
tragedy, *see* drama
trance, 23-5
  denial, 25
  manipulation, 25
triads, 36-7, 39-40, 42-3, 163 (*see also* threes)

UNCONSCIOUS, 7, 14, 18-19, 22-4, 32-4, 37, 43, 47, 50-52, 87-8, 90, 118 n, 123, 141-7, 150-51, 156, 157-8, 180, 181 n, 198

VAGINA, 65, 176
vampire, 15, 66, 149 n
variety, 28, 37, 56, 88, 90, 117-22, 163, 173
vegetarianism, 180, 202
volition, *see* will

WAR, 24-5, 35, 54, 61-2, 63, 65, 66 n, 69, 89 n, 118, 131, 172, 174, 177-8, 179, 182, 205
West, 34, 43, 51, 59, 69, 157, 159, 160 n, 171, 198 n
wife-beating, 175
will, 23, 105, 111, 112-4, 143 n, 146, 151-2, 184
wish, 24, 108
witch, 30, 38, 71
woman, *see* female
Women's Liberation, 176
Woodworth, R. S., 66 n, 142, 200
word, 93-5, 125-31, 184-5
Wordsworth, W., 156
workers, 54-5, 70-71, 154-5